THE INVISIBLE COLLEGE

Robert Lomas gained a first-class honours degree in electrical engineering before earning a PhD for research into solid state physics and crystalline structures. He went on to work on Cruise missiles and Fire Service Command and Control systems. He has always had a keen interest in the history of science and wrote an acclaimed biography of Nikola Tesla entitled *The Man Who Invented the Twentieth Century*. In 1986 Robert became a Freemason and quickly became a popular lecturer on Masonic history before co-authoring the international bestsellers, *The Hiram Key* and *Uriel's Machine*. He currently lectures in Information Systems at Bradford University School of Management. More information is available at www.robertlomas.com.

Also by Robert Lomas

The Man Who Invented the Twentieth Century

With Christopher Knight

The Hiram Key
The Second Messiah
Uriel's Machine

The Invisible College

The Royal Society, Freemasonry and the Birth of Modern Science

Robert Lomas

headline

First Published in 2002
by HEADLINE BOOK PUBLISHING

First published in paperback in 2003
by HEADLINE BOOK PUBLISHING

10 9 8 7 6 5 4 3 2

ISBN 0 7472 3977 0

Typeset in ACaslon by
Letterpart Limited, Reigate, Surrey

Printed and bound in Great Britain by
Clays Ltd, St Ives plc

Designed by
Anthony Cohen

HEADLINE BOOK PUBLISHING
A division of Hodder Headline
338 Euston Road
LONDON NW1 3BH

www.headline.co.uk
www.hodderheadline.com

Jointly dedicated to the memory of
Bro Sir Robert Moray
Freemason, scientist and spy, and
Dr John Hywel Roberts who inspired me
with a love of history

Contents

One of the more unusual sources of reference used in this research is William Preston's famous book, *Illustrations of Masonry* which was first published in London in 1772. This book is difficult for Masons to obtain, and virtually impossible for non-Masons, so to help readers interested in inspecting the original material, the text has been made available as a web-book at http://www.robertlomas.com. For full details of how to obtain your own usercode to allow you to access ***e-Illustrations of Masonry*** see page 383.

Acknowledgements

THIS BOOK WOULD never have been written if not for a casual conversation I had with Norman Macleod, the Assistant Grand Secretary to the Grand Lodge of Antient Free and Accepted Masons of Scotland, during which he asked me the test question, 'Who was the first Freemason made on English soil?' Fortunately, I knew the correct answer, 'Sir Robert Moray' and, reassured, Norman went on to tell me a lot about the importance of Moray's role in the Royal Society and talked eloquently about Sir Robert's many aspects as scientist, courtier, spy and Freemason. I would like to thank Bro Norman, his colleagues at Scottish Grand Lodge and Robert Cooper the Grand Lodge Librarian, for the help and encouragement they have given me during the research for this book.

I would also like to thank Jenny Finder and her Library staff at the Bradford University School of Management for their continuing assistance and cheerful acceptance of strange and unusual requests of long-lost books.

The structure and shape of the book emerged from a series of interesting discussions with my editor, Doug Young of Headline, and my agent, Bill Hamilton of A M Heath, who helped me grasp the importance and far-reaching consequences of the changes which the Royal Society set in motion and turned my vague idea into a firm project.

I would also like to thank Jo Roberts-Miller, Ian Marshall, Alan Butler, Chris Turner and Hugh Morgan for their help in the long process of editing and preparing the book for publication.

Finally, I would like to thank my wife and children for their on-going encouragement.

PROLOGUE

A Cold Winter's Day in Late November

One of the beautiful things about physics is its ongoing quest to find simple rules that describe the behaviour of very small simple objects. Once found, these rules can often be scaled up to describe the behaviour of monumental systems in the real world.[1]

Tim Berners-Lee, 1999

THE TALL, GAUNT man walked over to the window and looked out. The previous night had been bitterly cold, and the lawns at the side of his apartment were white with hoar frost. Using what warmth there was in his coarse soldier's hands he melted the ice from inside the casement and looked out east, across the privy gardens towards the cluttered roofs of the palace. The dull grey glow of the low winter sun, as it struggled to climb above the horizon, was giving little hope of anything but an overcast winter's day. Still, he thought, perhaps the rain would hold off a little while.

He wasn't in bad shape for a 52-year-old; he tried to keep himself fit and active. He looked towards the Sun dial, which he had recently constructed for the king in the centre of the lawns, but the light was too diffuse for it to cast any shadow. As he stood shivering in the winter's chill of that November morning he thought about the meeting he had arranged for the coming afternoon. Was he taking too great a risk in bringing together this group of men who had been sworn enemies for so long? Would the bonds of a single common interest be enough to persuade these men, who had all suffered so much during the recent war, to sit down together and talk? Was he hoping for too much in trying to persuade them to work together in harmony, to support the newly restored king? He shook his head to clear his thoughts.

A good soldier prepares his battle strategy before the onslaught begins and this man was a good soldier. He knew that it is better to capture an enemy army entire, rather than destroy it. He was aware that to fight, to conquer and then to destroy his old enemies would not help him to achieve his aims. He needed to break their resistance without fighting. Now was the time to apply the hard lessons learned during the years he had spent as a Quartermaster-General, civil engineer and spy in the armies of Scotland and France. He not only had to persuade his long-time antagonists to work with him, but he somehow had to make them believe that it was their own idea to do so.

How could he do this? Perhaps he could persuade one of the more extreme members of the opposition to chair the meeting? Who, he wondered, had the most to gain? Certainly the man who had lost the most was Wilkins. He remembered overhearing that garrulous young clerk of Lord Montagu's prattling on, earlier in the week. He was telling how Wilkins, the deposed Master of Trinity College and once favoured brother-in-law to Cromwell himself, was now reduced to preaching for coppers! This ex-Warden of Wadham College was struggling to live, crammed into the squalid lodging of yet another deposed cleric; and was reduced to acting as a chaplain for the penny-pinching lawyers of Grays Inn. Wilkins presented such a sorry spectacle that he was beginning to attract voyeurs to the Temple church, just to marvel at the extent to which the family of the late Lord Protector could be humiliated.

Yes, Wilkins would be flattered to be asked to chair the meeting, indeed if it was put to him in the right way he would accept it as nothing less than his right. That was the way to present it. Play to the man's vanity. Diplomatic skills learned in the service of the French had their uses, even in the uncertain world of Restoration England.

The clatter of horses' hooves and the rattle of a carriage stopping in the gateway, just below his rooms, drew him back from his reverie. Enough planning and scheming! The time for action had come. The king himself was forewarned; the king's supporters would already be setting off towards Bishopsgate; all that was left to do was to persuade able men, almost destroyed by the bloody events of the Civil War, to work with him. He felt his mouth go dry at the anticipation of the task. But if Britain was to survive the threat from the Dutch, he must succeed. He took a deep breath and turned away from the window.

Sir Robert Moray knew what had to be done and he knew how he intended to do it. He dressed carefully, donning the sombre black clothes he had favoured since the death of his wife. Was it almost ten years since she had passed away? He set off across the privy lawns towards the stone steps that led down to the Thames. Catching a sculler by the riverside he paid his sixpence to be ferried up river, almost to the Tower. There he disembarked, to walk up through the narrow, cluttered, reeking streets of Bishopsgate to the quiet haven of Gresham College.

After listening to a lecture on Astronomy from Christopher Wren, Sir Robert Moray went back to the rooms of Laurence Rooke, to the meeting he had been thinking about for so long. The day was Wednesday 28 November 1660. I don't really know what he thought and felt that cold November morning, but I know the results. That afternoon he created modern science!

Scientific Order from Political Chaos

So far this introduction has been pure speculation, but it is speculation based on fact. The man I have just described is a lost hero of science! He is responsible for the remarkable development in scientific innovation that has taken place over the last four hundred years and this book is the story of my personal quest to understand what he did and why he did it.

As a young scientist I learned that one of the highest honours to which a member of the scientific community could aspire was to become a Fellow of the Royal Society (FRS). This is the oldest and most respected scientific society in the world, its early members' names living on amid the physics I was studying. Looking down the list of early members was like reading the index of a textbook – Hooke's Law, Boyle's Law, Huygens's construction, Newton's Laws, Leibniz's theorem, Brownian motion; and this is ignoring lesser scientists such as Christopher Wren, John Evelyn, John Wilkins, Elias Ashmole, John Flamsteed and Edmund Halley.

But the men who founded this society were not just the first scientists; they were also the last sorcerers. Ashmole actually belonged to a society of Rosicrucians and was a practising astrologer; Newton studied and wrote about the Rosicrucian concepts of alchemy; while Hooke carried out magical experiments involving spiders and unicorn's horns.

The Rosicrucians, who took their name from their symbol of a Rosy Cross, taught about the magical harmony of the spheres which indirectly affected the harmony of the world. This ethereal music emitted strange, unseen, cosmic forces that affected the destiny of humans, and its consequences could be foretold from the positions of the stars in the heavens. Rosicrucians also believed that fire could be used as a universal means of analysis of the nature of matter and, in the right circumstances, could turn base metals into gold. However, they also claimed to be able to hold conversations with demons and angels! Not many of today's leading scientists would admit to such pursuits.

So what inspired an unlikely group of refugees from both sides of the Civil War to meet, form the world's oldest and most respected scientific society, and then go on to develop the tools of modern science? This was the question which started me off on this quest to understand how the Royal Society came to be formed. I wanted to know where this mixture of clergymen and politicians got the idea of forbidding the discussion of religion and politics at their meetings. In an age dominated by politics and religion it seemed a weird thing to do.

I couldn't avoid confronting these questions when I first read about the puzzling circumstances of this world-changing event. With the hindsight inbred by a scientific education it seems inevitable that the logic of science should succeed in banishing myth and superstition. In 1660, however, this outcome was not so certain. Was it just good fortune that brought so many important fathers of modern science together at this difficult time and inspired them to develop a new positive logic?

Only five months after Charles II returned to the throne of England this meeting of twelve men kick-started modern science. They soon started calling themselves the Royal Society. For scientific method to develop out of a community that believed in magic is an unlikely event in itself but when you add into the mix the fact that equal numbers of the twelve founder members of this Royal Society had been on opposite sides in the brutal English Civil War, such a fortuitous chance meeting of like minds seems not just improbable but impossible.

In the history of ideas there is usually a path which can be followed backwards, showing where ideas first appear and how they develop. However, if the traditional accounts of the formation of the Royal Society are to be believed, the concept of experimental science was developed, and fully formed, independently but simultaneously, on both sides during the Civil

War. Then, believe it or not, through a common interest in public lectures, all the members of the two groups just happened to meet for tea in London on a misty November afternoon. The rest, of course, is history. Here is how historian Sir Henry Lyons reported it:

> *Three centuries ago at the time of the civil wars a small group of learned men, who were interested in the Experimental, or New Philosophy as it was then called, made it their practice soon after 1640 to meet occasionally in London for talk and discussion at the lodgings of one of their number, or at a tavern conveniently near Gresham College where they often attended the professor's lectures . . . On the Restoration of the monarchy in 1660 those who were in London resumed their meetings that had been discontinued in 1658, and others who had been at Oxford joined them; by the end of the year they and a number of their friends having similar interests resolved to constitute themselves a Society of Philosophers, which they succeeded in doing.*[2]

The survivors of a civil war do not seem the most likely people to start up a new science club. Imagine that you had just survived living in Kosovo during the NATO/Serbian war of 1999. How likely do you think you would have been to institute a weekly dining club to discuss the esoteric aspects of astrophysics and would you have invited the son of Slobodan Milosevich to become your patron? Would you have been happy to pay a joining fee of about £1000 and a refresher fee of £100 a week to aid with the creation of visual aids for your weekly meeting? Would you have insisted on inviting members of the opposing side of the conflict to join you in these meetings? Or is it more likely that you would have been concerned to a greater degree with preserving your own and your family's safety? Perhaps you would have been too busy trying to keep some sort of income and to safeguard your property. Yet it is in just these sort of circumstances that the Royal Society of London was formed.

England was in the aftermath of a bitter civil war in 1660. After the death of Oliver Cromwell the country had tottered on the brink of fresh conflict, until the controversial decision was taken to invite the king to return. He had been forced, however, to promise to behave himself! Yet in this chaotic atmosphere of Restoration London the Royal Society was formed. It had an extremely high joining fee and a hefty weekly refresh fee, to be paid whether or not members attended.

Sir Henry's quotation paints a delightfully romantic picture of a group of gentleman scientists casually meeting for dinner and discussion while one of the most bloody periods of English history rages unheeded around them. During the war sons had been fighting their fathers; brothers had been trying to kill each other; great estates had been despoiled; a king had been publicly beheaded; and royal princes had fled to exile. For twelve years the country had been run on the personal whims of a military dictator and only the threat of another civil war had persuaded Parliament to restore the king. Yet, like an eye of calm in the midst of these furious storms, we are supposed to accept that these learned men had sat, calmly chatting about how to develop a radical new philosophy of experimental science. Only the perfect vision of hindsight can make this seem a natural way to behave.

When Sir Henry Lyons wrote his definitive history of the Royal Society in 1944 he was concerned with recording *what* had happened. This he did in an exemplary manner, but he didn't ask the question that has interested me since I first read his account of the circumstances of the foundation of the Royal Society. *Why* was it created?

Its founders questioned most of the basic premises of religion and theology of the time. Yet they managed to avoid conflict with extreme fanatics who were forcing their views on everybody else. Having successfully avoided the notice of the Covenantors, the Levellers, the Fifth Monarchists, the Papists and the followers of the *Book of Common Prayer*, they were still able to investigate such heretical matters as the practicality of witchcraft – and nobody challenged them!

Historian Arthur Bryant credits Charles II with a zeal for experimental science that led him to inspire the Royal Society:

> *With the return of the King, who had little use himself for abstract religious formulas, and preferred to test everything by his own keen commonsense, the new generation came into its own. Shortly after the Restoration, the Royal Society was founded in Gresham College, and the King became its first patron. When its members placed a spider in the midst of a circle of unicorn's horn, and the insect, disregarding the hallowed beliefs of centuries, 'walked out' – as the Society's minutes briefly record – something momentous happened.*[3]

This account of the king's personal interest is charming but highly unlikely. The experiment Bryant describes was an important step away from magic towards modern science, but the king was never the major driving force towards commonsense that Bryant implies.

It was no small feat for the founders of the Royal Society to develop a questioning, scientific philosophy 'at the time of the civil wars'. To constrain a spider using the horn of a mythical beast (in reality the horn was that of a rhinoceros) was to practise witchcraft and was flouting convention. During the rule of the Long Parliament (1645–7) just twenty-three years earlier, Matthew Hopkins, the Witchfinder General, executed 200 old women who were said to be practising witchcraft. In the seventeenth century magic and miracles were part of everyday life. Witchcraft was an acceptable explanation of ill fortune. Historian George Trevelyan writes that before the Restoration 'it would have been difficult to find more than a handful of men who openly avowed disbelief in the miraculous sanctions of the Christian faith, in one or other of its forms.'[4]

Yet by 1667, Bishop Thomas Spratt, the official historian of the Royal Society, speaking on behalf of its founders, wrote that the ancient miracles of bible times were privileged phenomena, unusual examples of God's interference with his creation, unlikely to be repeated. 'The course of things goes quietly along in its own true channel of natural causes and effects,' he wrote.[5]

The founders of the Royal Society seemed to avoid the problems of faith by accepting the Church's view on God and the soul, but questioning everything else. But, if the Royal Society's founders had been developing questioning views during the time of Matthew Hopkins, they must have kept quiet about them or they would have been persecuted. Yet for these ideas to appear fully formed, in 1660, they must have been around for a considerable time. By then the members of the Royal Society were giving no credence to witchcraft and were publicly laughing at the 'Popish miracles', as evidence of superstitious belief.[6]

Why had nobody noticed these ideas developing? Why, within the first few weeks of the Restoration, did science suddenly break free of the stifling dogma of religious belief and the repressive superstition of magic, and never look back?

The importance of this change in attitude should not be underrated. In the seventeenth century religion was undergoing a revolution. For the previous thirteen hundred years the Church had been systematically building an imperial faith, loosely related to the teachings of Christ and strongly supported by a verbal theology. To preserve its power the Church had to protect its theology. To keep their dogma intact and pure, Churchmen were extensively trained in methods of argument and disputation known as logical deduction. The Church controlled all the existing Universities and so set the agenda for education. This can be seen in the Church's treatment of Galileo's famous gravity experiment, said to have been conducted from the Leaning Tower of Pisa. His results showed that bodies of different weights fell at the same speed but his conclusion was logically disproved by the negative 'thinkers' of the Inquisition. Using theology to disprove experimental observation is something we find difficult to understand today, but that is because our whole basis of thinking about physical events was changed during the seventeenth century.

The change came about because this group of men met in London and decided to set up a society to study the mechanisms of nature. To make sure they were not distracted by dogma they forbade the discussion of religion and politics at their meetings. From this group, modern experimental science grew.

There had to be more to this story than the superficial record revealed, and so it proved to be. This book tells the story of my own quest to discover the political, economic and religious background to the formation of the Royal Society and how, in the process, I uncovered the hidden motives of one man, Sir Robert Moray. But before I could hope to understand the impact of the Royal Society I needed to look at the status of science before that fateful meeting of 28 November 1660.

CHAPTER 1

A World before Science

'A new and unprecedentedly effective form of knowledge and way of doing things appeared suddenly in Europe about 400 years ago. This is what we now know as science.'[1] Bryan Appleyard

SCIENCE IS NOT COMMON SENSE. Your eyes tell you that a chair is a solid object that you can safely sit on, but science tells you that the material of the chair is made up of many small parts with spaces in-between them. You could fall through these spaces! Yet you sit on the chair and it feels just as solid as it looked; but you still believe the scientist who explains that it is mainly empty space lightly sprinkled with atoms, even though the atoms are far too small for you to ever see them.

If you stand on a pebbly seashore, idly tossing stones out to sea, you expect to see your missiles fall in a curve into the near distance before plopping satisfyingly into the waves. You don't expect the stones to fly off in a dead straight line and disappear towards the horizon. But science tells you that any object continues to move in a straight line with unchanging speed unless a force acts upon it. Unless you are an astronaut you have never seen this happen and yet you believe it to be true.

If you were stopped in the street by a stranger offering to turn all your base metal into gold you would be suspicious and might think you were being tricked. Yet when British Nuclear Fuels turns chunks of uranium into weapons grade plutonium you accept the miracle as an everyday event.

In each of these three examples we are prepared to believe in things quite different from what we observe happening around us. We do this because we have been brought up in a society that accepts scientific explanations of the world.

Science is a way of thinking which not only explains events that have been observed but also predicts new facts that may have been undreamed of. In 1687 Sir Isaac Newton put forward a new theory of gravity. This theory is still in use today, for example, in working out the orbits of the satellites that bring us our television signals. But when he first published his ideas Newton contradicted two of the current theories about comets. The more popular one was that comets are a signal from an angry God warning that He will strike sinners and bring disaster. A less popular, and less dramatic, idea had been put forward by Johannes Kepler. He said that comets are celestial bodies that move in straight lines across the heavens. Now according to Newton's theory, some comets move in hyperbolas or parabolas, never to return, while others move in ordinary ellipses and appear again and again. At the time this was an incredible idea. However, the Astronomer Royal, Edmund Halley, used Newton's theory to observe a comet in the sky and predict to the minute when it would return, seventy-two years later. Right on time it did return and Halley's Comet has been eagerly watched by succeeding generations ever since. Halley's successful prediction did much to encourage the use of science as a way of thinking about the world.

Until the sixteenth century people believed magic was the

way to explain how the world worked. Queen Elizabeth I had a court magician called John Dee. Dee first came to the notice of Elizabeth's elder sister Queen Mary, when he tried to bewitch her and she, in turn, imprisoned him. When Dee was freed he took up with another alchemist Edward Kelly. They travelled about Europe, indulging in a bit of wife swapping and seeking an elixir of eternal life. Dee finally claimed to have invented this elixir soon after he returned to England, in a state of poverty.

His method of developing the magical liquid was very different from the way in which a modern pharmacologist works. Dee claimed to work with an angel by the name of Uriel, who was privy to all the God's knowledge of the world. Now such a responsibility is not normally given to just anyone and so Dee had to persuade the angel to part with his knowledge. He used the following incantation to summon the angel. (I suggest you don't try this spell at home, just in case it still works!)

> *Facilius Sine Comparatiorne a Deo impetrandum foret, vel a bonis spiritbus, quicuid homini utile reputare*

The angel spoke his own language, which he taught Dee to read and write. As well as giving Dr Dee the recipe for the elixir of eternal life the angel also predicted that Britain would have a vast empire. Dee recorded these conversations in various manuscripts.

Dee also carried out magical levitation displays using an obsidian stone which came from South America and a conjuring table which was engraved with the Enochian alphabet, used by the angels. These artefacts are now in the British Museum but, sadly, they seem to have lost their magical powers!

Despite his strange choice of research colleagues and his predilection for conjuring tricks Dee was also quite a competent mathematician. He secured his position as court philosopher at the court of Elizabeth when he revealed himself to the new queen as a master of electional astrology. Using the magical knowledge he had acquired from his conversations with angels he convinced the princess that he could calculate the most fortuitous date for her coronation, a date when the stars would favour her reign. He later advised the queen against adopting the Gregorian calendar, on the basis of complex calculations.

Dee's mathematical methods and astronomical observations were at the cutting edge of Elizabethan technology, yet he was a firm believer in magic. However, Dee outlived Elizabeth, so obviously his elixir didn't work for her. Elizabeth, the Virgin Queen, left no direct heirs and her crown passed to the line of the Stuarts. The new king, James I, sent Dee packing. Dee died soon afterwards, and remains dead to this very day, yet another sad victim of the failure of his elixir of eternal life!

But, during the reigns of the Stuart kings magic also died and science took its place.

Winston Churchill said of James, the first Stuart king of England:

> *He came to England with a closed mind and a weakness for lecturing. England was secure, free to attend to her own concerns, and a powerful class was now eager to take a hand in their management. Who was to have the last word in the matter of taxation? Was the king beneath the law or was he not? And who was to say what the law was? The greater part of the seventeenth century was to be spent in trying to find answers to such questions.*[2]

This questioning process involved civil war and regicide before answers were found, but in the midst of the battles between king and Parliament we are expected to believe that modern science suddenly popped up. No reason is given as to why this should be. Why a country which burned alive at least 100 elderly women a year, on suspicion that they were causing disease by casting the 'evil eye', should spontaneously develop a critical mass of discerning scientists is never questioned.

The old belief in magical forces did not die instantly, not even among the founders of the Royal Society; we find it still alive and well during the Civil War. In 1657 when Christopher Wren, later to become a founder of the Society, gave his inaugural lecture as Professor of Astronomy at Gresham College, he spoke of how London was particularly favoured by the 'various celestial influences of the different planets; both as the seat of the mechanical arts and trade and the liberal sciences'. No modern astronomer would dream of suggesting that the planets were capable of celestial influences, let alone that they might influence the future prospects of a city and its sciences!

Even the king, who in his spare time had become the patron of modern science, thought it perfectly normal to pay an astrologer to cast a horoscope for the best time to lay the foundation stone of the New Saint Paul's, after the Fire of London! Yet it was at this time that science began.

The Eternal Sky, Religion and Knowledge

The word science comes from the Latin word for knowledge, *scientia*. Modern science has two main functions, it enables us to know things and it enables us to do things. The success of science, as a means of searching for truth, is judged on how well it enables us to cope with our environment, and modern science has been very successful in improving our standard of living. We now judge science by how well it solves our problems, because the ground rules moved in the fifteenth century. Before the Royal Society changed our worldview, however, philosophers thought that a statement was real knowledge if enough people maintained a strong enough belief in it.

In 1589, when Galileo performed his famous experiment to see if heavy objects fell faster than light objects, his results showed that both weights fell equally fast. These results, however, were denounced by the Inquisition on the grounds of dogma, without any supporting evidence. The belief system that allowed this to happen sprang from an unholy alliance between the wisdom of Aristotle and the Church's assertion that it possessed the great Truth of Salvation. The Church's Truth said that God had made the world for the benefit of man and had sent His own son to ensure men understood His message.

But what had Aristotle, a Greek philosopher who had been dead for hundreds of years before the birth of Jesus, got to do with Christian Truth? This question puzzled me so I decided to investigate it by looking at the way in which this worldview, held by many people in the sixteenth century, had developed.

Before the seventeenth century people believed that the earth was the centre of the universe; that the sun, the stars, and the planets moved in circles around it; that the stars were made from an imperishable celestial fire; that they were arrayed throughout the universe on great transparent crystal spheres; and that the world had been created at precisely half past four on the afternoon of Thursday 22 October 4004 BCE.

Aristotle comes into this story, even though he died in 332 BCE, partly because he was such a popular and prolific writer. Many copies of his writings survived the collapse of the Greek empire, and were collected in Alexandria by Ptolemy. These writings took on an important role in Arab culture and eventually were brought back to Europe by the Crusaders. So Aristotle's writings became the basis of a medieval rediscovery of classical learning.

Aristotle presented his views on science in two books, *Physics* and *On the Heavens*. His ideas are based on two important phenomena: the movement of animals and the movement of the heavenly bodies. Living things move but dead things do not. It follows, therefore, that all movement is the result of the action of Will, either animal or Divine. This led him to a worldview dominated by causes and purposes. All things above the Moon are incorruptible and eternal; while all things below are subject to generation and decay. The earth is the centre of the universe and is made up of four elements: fire, earth, air and water. The stars, revolving in pure circles on their crystal spheres are made up of a fifth heavenly element. Everything has a purpose and it is the task of the philosopher to discover these purposes.[3]

Aristotle, no matter how much classical dons liked to praise him for the accuracy of observation, cannot always to be trusted implicitly. He insisted that women have fewer teeth than men. He married twice but looking into his wives' mouths and counting their teeth never seemed to have occurred to him. Perhaps he was afraid they might bite him! They would have had good cause because he also believed that children are healthier if they are conceived when the wind is in the North. (I have a whimsical image of him sending successive Mrs Aristotles out from the marital bed to look at the weather vane whenever they cuddled too close to him. Or perhaps he didn't trust their base instincts and went out to check for himself (which in those days of naked sleeping, leads me to worry whether or not his preparedness for the act of conception survived the ravages of exposure to the cold north wind). Some more of his outlandish claims are that the bite of a pregnant shrewmouse is dangerous to horses; that insomniac elephants can be sent to sleep by rubbing their shoulders with salt, olive-oil, and warm water; and that a man bitten by a mad dog will not go mad, but any other animal will.

Aristotle had many failings as an impartial observer but Christianity had no better way to explain the nature of the stars. Christian myth simply could not match the self-consistent and logical framework of Aristotle's splendid cosmology. Not until 1266 that is. In that year Thomas Aquinas wrote his theological Great Theory of Everything. In *Summa Theologiae* he accepted and improved on Aristotle's theories. Aquinas, however, insisted that the cause of the universe is God's intention that man should be saved from sin and hell. God had carried out this aim by the personal intervention of His Son, Jesus Christ. Aquinas concluded that knowledge of the world could only be an expression of the knowledge of the love and infinite wisdom of God. No mere human could question this Divine Will since its truth was not based on human confirmation but on the very authority of God. An attitude like this does not encourage casual questions and any inconvenient facts that do not fit into the Church's view of this world must be discarded. It follows that if the facts don't fit this theory then it must be the facts that are in error, because God cannot be to blame! This was how the church justified its treatment of Galileo!

The Persecution of Galileo

In 1633, Galileo (arguably the inventor of the astronomical telescope) fell foul of theology. He was summoned to Rome and forced by the Inquisition to make the following public statement:

> *I, Galileo, kneeling before you, most Eminent and Reverend Lord Cardinals Inquisitors-General against heretical depravity throughout the whole Christian Republic, having before my eyes and touching with my hands the Holy Gospels, swear that I have always believed, do believe, and by God's help will in the future believe, all that is held, preached, taught by the Holy Catholic and Apostolic Church. But whereas after an injunction had been judicially intimated to me by the Holy Office to the effect that I must altogether abandon the false opinion that the sun is the centre of the world and immovable, and that the earth is not the centre of the world and moves, and that I must not hold, defend, or teach in any way whatsoever, verbally or in writing, the said false doctrine.*[4]

The Inquisitors-General of the Holy Catholic and Apostolic

Church had total blind faith. They knew that the earth sat immovably at the centre of their universe; and this mistaken faith held firm against a mass of contrary evidence from their own observatories.

The Church, following the Jewish tradition, defines Easter as the first Sunday, after the first full moon, after the spring equinox. To plan their ecclesiastical calendars the leaders of the Church need to be able to predict the dates of full moons many years ahead. Now the cycle of the moon is unrelated to the solar calendar, which makes Easter a movable feast that can vary from 21 March to 23 April, dependent on the phase of the moon and exact day of the vernal equinox.

In the fifteenth century Cosimo de Medici, the ruler of Florence, commissioned a rising young draughtsman called Egnatio Danti to build a solar observatory to help predict the date of Easter. It consisted of a hole in the roof of a dome that would focus the sun's rays onto a scale on the floor of the building. By reading the scale it would be possible to know the exact day of the equinoxes. These solar observatories, know as *meridana*, were built into many churches, the most famous being within the dome of St Peter's Basilica in Rome.[5]

Systematic observations of the path of the sun's rays with these *meridana* soon showed that the sun did not revolve around the earth in a perfectly circular orbit, as Aristotle and Aquinas had taught. It could be clearly seen that the shape of the sun's image changed with the seasons, from a circle to an ellipse. During the winter months, when the sun was low in the sky the sunlight hit the hole at a different angle from rays of the higher summer sun. The dogma of the Church said the earth was fixed and the sun rotated about it in a perfect circle, since God could not make anything that was less than perfect. For this to be true the shape of the sun's image projected by the ecclesiastical *meridana* should have remained a perfect circle. It didn't!

But that wasn't all! The priestly keepers of these observatories also noticed that the angle between the earth and the sun changed slightly with time. (We call this phenomena precession and it is caused by a very slight wobble in the axis of the earth.) If you follow the logic of Galileo's forced confession there is only one acceptable explanation for this behaviour. To support their faith the seventeenth-century Church, and its Inquisitors-General, publicly professed a belief that the entire Universe danced and turned about the immovable throne of the Pope!

Giovanni Cassini was an astronomer in the seventeenth century. He found several new moons going round Saturn while working for his immovable Pope. By combining the data from St Peter's *meridana* with his own telescopic observations he was able to predict accurate orbits for Mars and Venus. He had used mathematical techniques that assumed the planets and the earth were moving in elliptical orbits about the sun. But he never ever dared to give an opinion on the earth's immobility. Perhaps he didn't want to have to make a public confession of his errors![6]

So, when modern science was born, the most relevant and powerful knowledge was knowledge about God, the Devil, Heaven and Hell. To make a mistake about these matters of faith involved punishment by eternal damnation. This threat naturally made theologians careful. Theological knowledge could not afford to be fallible; it must be beyond doubt. In this intellectual environment the Royal Society took root. Not the most favourable climate to encourage an open, questioning attitude towards the workings of nature!

Revolutionary Ideas

Science grew out of superstition and magic but as it gave birth to technologies, it assumed a much greater political importance. The strength of the Tudor monarchs was based on the technology of artillery and the use of gunpowder. The invention of the mariner's compass enabled ships to navigate to the New World of the Americas. From that time onwards the main interest that most rulers have shown in science is how it could be used to increase the power of their weapons of war, or improve the strength of their military forces.

The year the Royal Society was born, religion was still an important issue in England. One of the main causes of the bitterness of the Civil War was the differences in doctrine between the two sides. Indeed, it was the disarray of the various religious factions that enabled General Monck to bring about the Restoration. Any form of fanaticism can lead to an intolerant society. If you are an intense believer in any religious idea you will be prepared to face martyrdom, you can live a happy life of great hardship and even enjoy a happy death if it comes quickly. You may inspire converts, create armies, promote hatred of any dogma that your cause does not accept and be immensely effective in promoting your beliefs, as well as suppressing any other viewpoints.

Scientist Richard Dawkins has said this about religious faith:

> *[Blind Faith is] . . . powerful enough to immunize people against all appeals to pity, to forgiveness, to decent human feeling. It even immunizes them against fear, if they honestly believe that a martyr's death will send them straight to heaven. What a weapon! Religious faith deserves a chapter to itself in the annals of war technology, on an even footing with the longbow, the war-horse, the tank and the hydrogen bomb.*[7]

Any fanatical creed does harm. This is most obvious when one set of fanatics competes in outrageous behaviour and hatred with another group. Bertrand Russell gave the example of a friend of his who was a fanatical supporter of an international language. This man preferred Ido to Esperanto and explained to the bemused Russell just how depraved and unimaginably wicked the speakers of Esperanto were by trying to promote it as an international language.

Often this hatred of competitors becomes the most important feature of a fanatical belief. Some people whose religious belief tells them they should love their neighbours as themselves, reserve the right to hate anybody who refuses to accept this view. This hatred arises from an attitude that accepts unquestioningly a belief on the basis of authority, without admitting any questions of why the belief should be held.

In 1660 the Roman Catholic Church had already held this type of fanatical view of the world for four hundred years. It had just demonstrated, by its treatment of Galileo, that it was not prepared to tolerate any deviation from its preferred truth. The Protestant Puritans of England had rejected some of the extreme dogma of the Roman Catholic Church and had instead sought their support from the Bible. The Protestants, however, having won this victory proceeded to persecute each other for small deviations in their interpretations of God's Will.

Probably the most extreme example of this attitude can be seen in Cromwell's abortive attempt to establish a Parliament of Saints in 1653. This was later known as the Barebone Parliament, after one of its more out-spoken members Praise-God Barebone. Following his violent dissolution of the Rump Parliament Cromwell called on the independent churches of each shire to nominate suitable candidates to act as members of parliament. He asked for 'persons fearing God and of approved fidelity and honesty'.[8] From the lists submitted, one hundred and fifty members were selected and on 4 July, summoned to Whitehall.

This experiment in government was a total disaster. The Barebone Parliament even attempted to abolish the Common Law and substitute the Law of Moses in its place. Cromwell said of this Parliament, 'Fain would I have my service accepted by my saints, if the Lord will, but it is not to be so. Being of different judgments, and those of each sort seeking most to propagate their own, that spirit of kindness this is to all, is hardly accepted of any.' In the end the crunch came when the Parliament tried to abolish the state-endowed Church, which Cromwell supported. Eventually the Barebone Parliament was forced to dissolve and the fighting rabble of religious fanatics disbanded.

With the perfect vision of hindsight it is clear that the most inspired rule that the founders of the Royal Society adopted in their meetings was to ban the discussion of religion and politics. Thus at a stroke they removed the two major subjects that would cause them to quarrel. In the circumstances of the time it must have seemed a strange and unnatural idea. Where did they get it from? To find out, I knew I would have to look more closely at the detailed circumstances of the commencement of the Royal Society.

The Traditional History of the Royal Society,

As I have already mentioned, Sir Henry Lyons, in the introduction to his history of the Royal Society, did not find it unusual that within six months of returning from exile, King Charles II was actively supporting the formation of a society which on the surface would seem to have been of little interest to him. Sir Henry also accepts the Royal support without comment, saying:

> *In December their project received the approval of King Charles II and the promise of his support, which was followed a few months later by his permission to use the title of the 'Royal Society'. From such small beginnings did the Society arise.*[9]

The Royal Society of London is the oldest and most successful club of experimental scientists in the world. We enjoy a high standard of living today only because the Royal Society changed public attitudes to science and technology.

In the months before the Society formed, England went through a period of great unrest. Indeed, it looked for a while as if another Civil War would begin, and yet the first meeting of this Society drew together senior figures from both sides of the conflict. This appeared to be a real puzzle as I struggled to understand what had been going on.

Surprisingly quickly the Society had attracted the attention of the newly restored king, who must have had many more pressing matters on his mind. Yet, within a week of its first meeting Charles II offered the Society his blessing, despite the fact that this first meeting was chaired by Oliver Cromwell's brother-in-law; and the king was well known for his hatred of the dead Lord Protector, who had murdered his father.

The Royal Society was born of twelve disparate and ill-assorted men, meeting on a cold November afternoon in the rooms of the Professor of Geometry at Gresham College in London. At first sight they seem to share nothing beyond a degree of wealth (to afford the fees) and a curiosity about the workings of nature. But as I started to investigate these men I soon discovered that they formed two clear groups; and these groups had no reason to even show any regard for each other, let alone to wish to meet for regular scientific tea parties. What is more, some of the individuals concerned were far from wealthy. These founding twelve differed in terms of politics, scientific expertise and social rank. They formed a very strange political mix indeed.[10]

I was very puzzled. What could have brought together these old enemies to establish, of all things, a scientific society? And what was so interesting about the ideas they discussed at that meeting which, over three hundred years later, are still inspiring the world in scientific debate? Were they in need of entertainment? Were they idle gentlemen of leisure, with nothing better to do with their time? Perhaps they were short of amusement in the evenings, after all television had not yet been invented. I really wanted to know what inspired them to change the world.

When I first began to study these men, I wondered if it would be possible to answer such questions, so long after the event. But, fortunately, from that first meeting, the society kept a journal book and its opening pages were at least able to tell me what its founders did, if not why they did it:

> *These persons following, according to the usual custom of most of them, mett together at Gresham College to heare Mr Wren's lecture, viz. The Lord Brouncker, Mr Boyle, Mr Bruce, Sir Robert Moray, Sir Paul Neile, Dr Wilkins, Dr Goddard. Dr Petty, Mr Ball, Mr Rooke, Mr Wren, Mr Hill. And after the lecture was ended, they did, according to the usual manner, withdrawe for mutuall converse. Where amongst other matters that were discoursed of, something was offered about a designe of founding a Colledge for the promoting of Physico-Mathematical Experimental Learning. And because they had there frequent meetings with one another, it was proposed that some course might be thought of, to improve this meeting to a more regular way of debating things, and according to the manner in other countryes, where they were voluntary associations of men in academies, for the advancement of various parts of learning, so they might doe something answerable here for the promoting of experimental philosophy.*
>
> *In order to which, it was agreed that this Company would continue their weekly meeting on Wednesday, at 3 of the clock in the tearme time, at Mr Rooke's chamber at Gresham College; in the vacation at Mr Ball's chamber in the Temple, and towards the defraying of occasional expenses, every one should, at his first admission, pay downe ten shillings and besides engage to pay one shilling weekly, whether present or absent, whilst he shall please to keep his relation to this Company. At this Meeting Dr Wilkins was appointed to the chaire, Mr Ball*

> *to be Treasurer, and Mr Croome, though absent, was named the Registrar.*
>
> *And to the end that they might be the better enabled to make a conjecture of how many the elected number of this Society should consist, therefore it was desired that a list might be taken of the names of such persons as were known to those present, whom they judged willing and fit to joyne with them in their designe, who, if they should desire it, might be admitted before any other.*[11]

This seemed simple enough. A group of gentlemen met up by accident as they regularly attended public lectures in London. They so much enjoyed talking about science that they set up a scientific society to amuse themselves. Most of them weren't short of money so they fixed a ten shilling joining fee and a shilling a week contribution to pay for their amusement (this would equate to about £1000 to join and an ongoing fee of £100 per week in today's terms). Not cheap entertainment at that time! And no mention made of the wildly different political backgrounds of these gentlemen.

Conclusion

Prior to the establishment of the Royal Society, science had been completely dominated by religion and suppressed by theological argument. The general climate was superstitious and most people believed in magic. The Church had a monopoly on thinking and was not swayed by facts, because it already knew God's Truth. Any experimenter who challenged the views of the Inquisitors-General was a heretic and was punished accordingly.

In the middle of the seventeenth century this attitude changed completely and from that time, modern science began to grow. The change occurred towards the end of one of the most bloody periods of British history, hardly a time to encourage philosophical contemplation. The people who were involved were drawn from both sides of the Civil War, and at first sight, seemed an unlikely group of people to be meeting socially, to amuse themselves by the study of experimental science. Yet these people started a wave of change that surged up to form our modern scientific society.

Who were these people? Where had they got their revolutionary ideas? These were the next questions I needed to think about. So I decided to start by collecting as much information as I could about each of these twelve founders.

CHAPTER 2

The Founders of the Royal Society

At the time of its formation, the Royal Society embodied a new philosophy and a new scientific attitude; and its prompt recognition by the restored monarchy of Charles II, which gave it its royal charter in 1662, showed a new attitude on the part of the monarchy. For not only was the patronage of scientific research by the Stuart monarchy something new in itself: it was also, in this instance, politically surprising.[1] H R Trevor-Roper

SO, WHO EXACTLY WERE THESE men who founded the Royal Society? Sir Henry Lyons had said they were all regular attendees at Gresham lectures so perhaps I would find evidence of this.

The Right Revd John Wilkins

The man in the chair of that first meeting was the Reverend John Wilkins. Wilkins was born in 1614 at Fawsley in Northamptonshire. He was the son of an Oxford goldsmith and the grandson of a country vicar, John Dodd. He went on to be a very successful churchman himself. By the time he died, in 1672, he was Bishop of Chester.

Wilkins was a remarkable survivor. During the build up to the Civil War, the young Wilkins became a great supporter of Parliament. He got his reward. On 12 April 1648, (after

Charles I's surrender to the Scots at Newark), he was made Warden of Wadham College, Oxford. The job was vacant because Parliament threw out the previous warden, for holding Royalist sympathies. Eleven years later Wilkins successfully sought a special ruling from the Lord Protector, Oliver Cromwell, that he might 'be relieved of the prohibition against marriage' that was a requirement of his post. As soon as this was granted he married Cromwell's sister Robina.

In all the books about Wilkins and the Royal Society there is hardly any mention of his wife. The few writers who refer to his marriage tacitly imply that he married for love and lived happily ever after. Typical is the only comment made by Dr E J Bowen about Wilkins' wife in a lengthy biographical account of the founders of the Royal Society:

> *In 1656 he married Robina, a sister of the Protector who gave him dispensation from the Statutes of the College which prohibited the Warden from marrying.*[2]

This brief mention leaves the reader to paint a romantic picture of a love-struck celibate academic seeking a change in the law, just so he can marry the woman of his dreams. But the marriage can hardly have been a love match because Robina was a widow of sixty-two years of age. Robina's previous husband had been Peter French, the canon of Christchurch Oxford. She had a daughter, Elizabeth French, who was almost the same age as Wilkins.

When her husband died she would have had been forced to leave the church house and so would have been in need of somewhere to live. Clearly Mrs French was available and well connected, obviously a fact that counted far more than her age, or sexual allure, when Wilkins came to assess her suitability as a prospective bride.

Wilkins himself was not the type of man to have ladies

continually swooning over him. He was a mature 42-year-old eccentric bachelor, given to practical joking and quite set in his ways. He was also rather fond of taking good dinners with his male friends. (John Evelyn wrote in his diary on 10 July 1654, 'supped at a magnificent entertainment in Wadham Hall invited by my excellent and dear friend Dr Wilkins, then warden'.)

Wilkins' writings show him to be a man of varied interests. He designed wind-driven chariots, wrote about ways of travelling to the moon and created a garden with a haunted statue. All this before he even thought about wooing the aging Robina. The statue had a hidden tube running from its mouth to a more distant part of the garden. When Wilkins was showing his friends, such as John Evelyn, around the garden he would drop back as they approached the statue. Hiding himself behind a bush he would then slip out of sight and make the statue speak via the tube. He was always much amused by his friend's consternation. Evelyn said of this statue 'He [Wilkins] had contrived a hollow statue which gave a voice and uttered words, by a long and concealed pipe which went to its mouth, whilst one spake thro it, at a good distance and which at first was very surprising.' Wilkins' talent for making a statue repeat sweet nothings to her must have really impressed the elderly Mrs Robina French. She married him just as soon as her brother gave his consent!

However, children and the delights of the marriage bed did not seem to be in the forefront of Wilkins' ambitions. John Evelyn had no doubt about his real motive for marrying Robina. On 10 January 1656 Evelyn wrote in his diary:

> *I heard Dr Wilkins preach in St Pauls before the Lord Mayor, showing how obedience was preferable to sacrifice etc. He was a most obliging person, but had married the Protector's sister, to preserve the Universities from the ignorant Sacrilegious*

> *Commander and soldiers, who would fain have been at demolishing all bothe places and persons that pretended to learning.*

The dating of Evelyn's comment suggests that immediately after Wilkins married Robina, the happy couple set off on the two days' coach journey from Oxford to London, to enable Wilkins to preach at St Paul's. Dr Wilkins' behaviour does not seem typical of a newly married man and he seems to have had a very small window of opportunity to consummate his new marriage in any degree of comfort. Was he really practising what he preached and sacrificing the conjugal delights of his new bride to be obedient to his calling of the Church?

The week after his sermon at St Paul's, Wilkins accompanied Evelyn on an expedition to visit Mr Barlow, a noted painter of birds and beasts. His new bride did not go with him.[3] Perhaps now she had a secure home in Oxford, Robina did not see the need to make any marital demands upon the long celibate Doctor. Indeed, Wilkins' new marriage did not seem to curtail in any way his regular nights out with the boys. On 12 April 1656 he joined Evelyn, Robert Boyle and a number of other gentlemen for dinner at Evelyn's house. Afterwards Wilkins and the party of gentlemen went on a trip up to London to visit Sir Paul Neile, who was reputed to be making high quality magnifying glasses. Wilkins stayed a month with Neile and was still a guest a month later when Evelyn next went up to London. The new Mrs Robina Wilkins was clearly not possessive about, or insistent upon clinging to, her new husband.

Whatever Wilkins' motive for getting wed, the match surely helped his career. One of Cromwell's last acts before dying was to order Parliament to appoint him Master of Trinity College, Cambridge. This was confirmed by Robina's nephew Richard Cromwell, who briefly became the Protector after his father's death.

Wilkins' plan for rapid preferment fell apart, however, when Charles II returned to the throne. The anonymous editor of Wilkins' *Mathematical and Philosophical Works*, published in 1708 said this of his fortunes in 1660:

> *After king Charles the IId's restoration, he was ejected from thence and became preacher to the honourable society of Gray's Inn, [lodging] in the room of Dr [Seth] Ward.*[4]

Samuel Pepys, another well-known diarist who was later to play an important role in the Royal Society, first met Wilkins on Sunday 25 November 1660. Pepys went to hear Wilkins preach at Gray's Inn Temple. He comments that he had gone to that church 'to hear the late master of Trinity College, who had been ejected from his post by the king'.[5] So when Wilkins chaired that fateful meeting on Wednesday 28 November he was in dire circumstances. He was an object of curiosity for the more literate men of London; he had resigned his Mastership; he was homeless; and he had been driven from his new job in Cambridge. Reduced to sharing the lodgings of Seth Ward, Wilkins must have been hard pressed to find the quite substantial subscription needed to join the new Society.

So, what could I find out about Wilkins' reluctant housemate, Seth Ward? He had had a very chequered career during the Civil War. Although he was a supporter of Parliament he managed to upset Cromwell. A year after Wilkins became Warden of Wadham College, Ward took up the Savilian Chair of Astronomy. Thus Wilkins and Ward became close friends and in May 1654 they wrote a book together. It was called *Vindiciae Academiarian.* It was a vigorous defence of the rights of the Ancient Universities to stay as the only providers of university education. They strongly defended the rights and privileges of Oxford and Cambridge, insisting that no others were necessary or desirable.[6]

Although Ward was elected Master of Jesus College, Oxford, he soon fell foul of Oliver Cromwell's ill will and was ejected from the mastership in 1657. Perhaps it is now clearer why John Wilkins decided to marry Cromwell's sister, soon after publishing *Vindiciae Academiarian.* If he hadn't become part of Cromwell's family, his attack on the Protector's higher education policy might well have done him much more damage.

Ward later co-wrote a condemnation of Thomas Hobbes, with mathematician John Wallis.[7] After Cromwell's death Ward became Master of Trinity College, Oxford, only to be ejected from his job by Charles II in August 1660. Ward was fortunate to obtain a living as a prependary priest in London[8] and was able to offer accommodation to Wilkins, his fellow refugee. Although Ward was a competent mathematician and astronomer, he was not at the 28 November meeting. Indeed, he did not become a fellow of the Royal Society until over a year later, on 18 December 1661, just before becoming Bishop of Exeter.[9] He was, however, among the 41 extra members nominated as potential fellows at the first meeting, so Wilkins had not forgotten his friend and benefactor.

The journal book of the Royal Society says that on 5 December 1660:

> *Sir Robert Moray brought in word from the Court, that the king had been acquainted with the designe of the meeting. And he did well approve of it, and would be ready to give encouragement to it.*[10]

This is the politically surprising event that Trevor-Roper referred to in his quotation that opens this chapter. Charles II had proved to be reasonably tolerant of what had happened during the war but one thing he had not been prepared to overlook was Cromwell's role in his father's death. A blunt statement in Microsoft's Encarta Encyclopaedia says it all:

> *After the restoration of Charles II in 1660, Cromwell's disinterred body was hanged as that of a traitor, his head put on a pole mounted above Westminster Hall, and his body buried at the foot of the gallows.*[11]

The king had borne such a strong grudge that he was prepared to mutilate the dead body of Oliver Cromwell. He had also removed Cromwell's brother-in-law from Trinity College, to replace him with an incumbent of his own choosing. Yet now that same king was freely offering support and encouragement to a speculative venture, chaired by the man he had so recently sacked from Cambridge. This did not make much sense to me.

Marriage to Robina had helped Wilkins when Cromwell had been in power; now his close links with the Cromwell family were no longer an asset. I discovered later that, fortunately for Wilkins, the elderly lady had conveniently died before he moved to London.

But what could I find out about those other 'regular attendees' at the scientific lectures held at Gresham College?

Viscount William Brouncker

Brouncker is also a puzzle. Sir Harold Hartley, another historian of the Royal Society, poses himself the question 'Why was he [Brouncker] chosen as the first President of the Royal Society rather than John Wilkins, John Wallis, Robert Boyle or Sir Robert Moray?'[12] He then avoids answering the question, simply applying hindsight to say, 'The wisdom of the choice was proved by the devoted and able service he gave to that high office during the infant years of the Society.'

John Evelyn, one of the members of the first Council of the Royal Society is explicit about why William Brouncker was created the President when he records in his diary for 20 August 1662:

I was this day admitted and then sworn one of the present Council of the Royal Society, being nominated in his Majesties Original Grant, to be of this Council, for the regulation of the Society, and making of such Laws and Statutes as were conducible to its establishment and progress: for which we now set a part of every Wednesday morning 'till they were all finished: My Lord Viscount Brouncker (that excellent Mathematician Etc) being also, by his Majestie, our Founder's nomination, our first President.

So the answer to Sir Harold's question is that Brouncker was the first President because the king insisted on having him in that post. This does, of course, pose a further question. Why did the king insist on having Brouncker in this position of authority? Charles even went so far in the charter of 22 April 1663 as to say:

we have assigned, nominated and constituted and made our very well-beloved and trusty William Viscount Brouncker to be and become the first and present president of the Royal Society.

Clearly, Charles had no doubts whom he wanted in charge of the Royal Society. I hoped that the reason for this choice would become clearer as my investigation unfolded, because as yet I couldn't see any reason for the choice.

Brouncker was a Royalist who kept his head down during the period of the Protectorate. He spent his time translating Descartes' theories about music into English. He was also a capable mathematician. Brouncker had studied under John Wallis, the Savilian Professor of Geometry at Oxford, who I knew was a friend of John Wilkins. As a signatory of the Declaration of 1660, Brouncker had played his part in the Restoration when he was returned as MP for Westbury in the Convention Parliament.

Brouncker wanted to be sure that the newly restored king knew of his loyalty, so he made Charles a gift of a small pleasure craft. He had designed this boat on radical new lines, and had it constructed by a famous shipwright of the time. He said he gave the king this gift 'to mark his restoration to the throne of England'.[13]

He became the first President of the Royal Society. Immediately afterwards he was made a commissioner for the navy. This was an important appointment because it was made at the time a naval war with the Dutch was looming. It was not an easy task, however. Brouncker was appointed to a navy suffering from corruption, lack of discipline and severe shortage of funds. Samuel Pepys became Clerk to the Navy Board. He and Brouncker worked for many years to improve matters.

I may not have discovered why he became the first president but I now knew why it had been easy for Brouncker to attend that first meeting. As a Member of Parliament it is not surprising that he happened to be in London on 28 November 1660. But who invited him? It would certainly not have been Wilkins as they were on quite different sides of the political fence. Brouncker had just recovered some political power while Wilkins was a discredited down and out.

The Right Honourable Robert Boyle

The next attendee listed is Robert Boyle. He was 33 years old when he went to that 28 November meeting. Boyle had spent most of the Civil War writing theological tracts in the depths of Dorset. During the early part of the Protectorate he moved to Ireland but in 1653 John Wilkins wrote to him, inviting him to come and settle in Oxford at Wadham College, where he could continue his studies of nature and science. Boyle moved to Oxford in 1654. He was reputed to be an extremely competent physicist and gave his name to the law that relates the pressure and the volume of a gas.

Boyle lived in Oxford until 1668 when he moved to London. If he was a regular attendee at the Wednesday afternoon lectures at Gresham College he must also have been a regular traveller. Gresham College in Bishopsgate, London was a 120-mile round trip from his home, next door to the Three Tuns public house in Oxford. With more than a day's ride each way he would have had little time left for anything else, so it seems safe to assume Robert Boyle did not make it his usual custom to attend the lectures on Wednesday afternoon. He did sometimes come up to London to stay with his sister in Chelsea, as John Evelyn visited him there on 7 September 1660. However, the lecture to be given by Christopher Wren must have attracted him enough to make the journey and somebody must have told him about the meeting to encourage him to come. Who might that have been? Wilkins is certainly a possible candidate for inviting Robert Boyle, as Boyle knew him well from his time at Oxford. Boyle's name was also on the list of people who were invited to join after the first meeting. But, as he was there when the meeting started why did the others need to write to him?

Alexander Bruce, Earl of Kincardine

Boyle was certainly not invited by the next man on the list, Alexander Bruce. Alexander was a Scotsman and the younger brother of Edward, the first Earl of Kincardine. Edward Bruce had been made an earl by Charles I in 1647. The Bruce family had supported the Stuarts throughout the Civil War and after Charles II's abortive attempt to drive out Cromwell, in 1650, Alexander had been forced to flee Britain and had settled in Bremen. He remained there until 1660, when hearing of the proposed restoration he had gone to The Hague to join Charles II for his return to London. He travelled back to London with Charles's entourage and set up house in Charing Cross. What I found out about Bruce is mainly gleaned from a long series of letters, some one hundred and twenty, written to him by Sir Robert Moray, between 1657 and 1673. The main scientific content of these letters concerns coal-mining and the construction of watches, both topics in which Bruce had an interest.

Bruce was described by Bishop Gilbert Burnet, who knew him well, as:

> *both the wisest and the worthiest of men that belonged to his country, and fit for governing any affairs but his own; which he by a wrong turn, and by his love for the public, neglected to his ruin; for they consisted much in works, coals, salt, and mines, required much care, and he was very capable of it, having gone far in mathematics and being a great master of mechanics. His thoughts went slow, and his words came much slower; but a deep judgement, appeared in everything he said or did.*[14]

Bruce's health was poor after his return from exile and he stayed in London recuperating until 1662. That year he succeeded to his brother's title and returned to live in Culross, Scotland. A series of Wednesday afternoon lectures on science sounds just the sort of thing to cheer him up, during his convalescence, so he is very reasonably considered a likely candidate for the role of 'regular attendee', at least after the Restoration. Or was it perhaps his close and long-time personal friend Sir Robert Moray who had invited him?

Sir Robert Moray

Sir Robert Moray is the next founder listed and was also a Scot. The date of Moray's birth is uncertain but it was 10 March, either 1608 or 1609. He was educated at St Andrews University and served with the Scots Guard of Louis XIII in 1633.[15] Towards the end of Cardinal Richelieu's life Moray became his favourite and historian Patrick Gordon said of his relationship to Richelieu:

> *Wherefore, choosing forth a man fit for his purpose amongst a great many Scots gentry that haunted the French court he chooses forth one, Robert Moray, a man endowed with sundry rare qualities, and a very able man for the Cardinal's project.*[16]

The 'Cardinal's project' was spying! In 1638 the General Assembly of the Covenantors in Scotland were rebelling against Charles I. The following May, Charles lost the First Bishop's War and had to make concessions to the Scots. Richelieu gave Moray a commission, promoting him to Lieutenant-Colonel in Louis's elite Scots Guard, and dispatched him to Scotland. Ostensibly he was supposed to recruit more Scots soldiers but he also admitted that he had the objective of assisting his fellow countrymen in their dispute with Charles, by causing trouble for England.

Moray was appointed quartermaster-general of the Covenantors' Army, in 1640. He would have been responsible for laying out camps and fortifications, where his knowledge of mathematics and surveying would have been extremely important. He marched south with the Scottish Army towards the Tyne and played his part in defeating the Earl of Stafford's English conscript Army at Newcastle.

After Charles I had agreed to the establishment of total Presbyterianism in Scotland, the king had been accused of complicity to kidnap the Marquis of Argyll. The Scots turned on him and insisted he paid for the ongoing support of their Army in Newcastle. By 1643 Moray was acting as a liaison officer between the Covenantors' Army and Charles I, in his court at Oxford. He must have been good at the job because on 10 January 1643, Charles knighted him. During this period, when he acted as a negotiator between Charles, the Scots and the French he seemed to have developed a close friendship with both the king and the Prince of Wales.

Soon afterwards Sir Robert returned to France and was promoted to Colonel in the Scots Guard. He had the misfortune to be captured by the Duke of Bavaria while leading his regiment into battle on 24 November 1643 and was imprisoned for eighteen months. During this time he studied magnetism, until he was freed on 28 April 1645 when the French decided to pay a ransom of £16,500 for him.[17]

Just before Moray had been ransomed and returned to Paris, Charles I had been defeated at Naseby. Cardinal Mazarin, Richelieu's successor, now sent Moray to London. He was made a member of the French Ambassadorial party whose job was to support the Scottish Commissioners, who had in their turn been appointed by the Edinburgh Parliament to negotiate with the king. So Moray became involved in the drawn-out and awkward negotiations between the defeated king and his victorious people.

When these talks broke down and Charles fled to Newark, to throw himself upon the mercy of the Covenantor Army, Moray again became closely involved with the king. On 24 December 1646, Moray arranged for Charles to escape to France. Sir Robert had paid for a vessel to be lying ready at Tynemouth. The king was being held in Newcastle. Bishop Gilbert Burnet said of the episode:

> *Sir Robert Moray was to have conveyed the king there [to Tynemouth] in a disguise; and it proceeded so far that the king put himself in the disguise and went down the back stairs with Sir R Moray. But His Majesty, apprehending it was scarce possible to pass through all the guards without being discovered, and judging it hugely indecent to be caught in such a condition, changed his resolution and went back.*[18]

Burnet claimed to have had the story direct from Sir Robert. The whole history of the Civil War might have changed if Moray had succeeded in his plan to get Charles to France. However, it was not to be. Once Charles was sold to Cromwell, for the price of the Covenantor Army's back pay, he was taken to London to be put on trial for treason. Moray, meanwhile, returned to France.

After the execution of Charles I, and at the request of the Earl of Lauderdale, Moray opened negotiations that led to Charles II going to Scotland to be crowned King of Scots, at Scoon [the modern spelling if you are looking for the town on a map is Scone] in 1650.[19] Charles's campaign, with a Scots' army, to recover England from Cromwell failed at the Battle of Dunbar and, after hiding for a while in an oak tree, Charles fled to France. Moray remained in Scotland.

Soon after Charles's flight Moray married Sophia Lindsey, the beautiful sister of the Earl of Balcarres. In July 1652 the newly married Morays returned to Edinburgh for the birth of their first child, and also to help organise a rising to restore Charles to the throne of England, but neither was to be. Sophia suffered a protracted and agonising labour before finally dying, on 2 January 1653, with the stillborn child. Once again the Scots were defeated by Cromwell, this time at the battle of Loch Garry in July 1654. Now Moray was accused of betraying the king but was cleared, after writing directly to the king and appealing his innocence. Moray returned to France, and would never again marry. In later life he was described as 'a single man, an abhorrer of women'.[20] Evidently no other woman could ever replace the gaping hole Sophia's death left in his personal life.

By 1655 Moray was back in Paris. At 46 he was getting too old for the Scots Guard. He resigned his commission and after spending a year in Bruges retired to Maastricht where he spent his time studying science and carrying out that protracted correspondence with Alexander Bruce. In September 1659 he went to Paris to meet with Charles and proceeded to take part in the negotiations with Monck to have Charles restored to the throne of England.

When the king returned to England, in late June 1660, Moray stayed on in Paris for some months. When he travelled to London, in August,[21] the king greeted him warmly. 'His Majesty received Robert Moray with crushing and shaking of his hand.'[22] Charles immediately found Sir Robert a grace and favour house within the grounds of the Palace of Whitehall. A drawing of Whitehall in 1680, held by the London Topographical Society, shows Sir Robert's quarters to be a small house situated just inside the Horse Guards Gate and looking out over the privy garden. The site of this house was exactly opposite where Dover House now stands on the present Whitehall.

It was from this house that Sir Robert set out to Gresham College on 28 November. He had been living in London for three months, having spent the previous ten years in exile. He could hardly have been a regular attendee of the Gresham meetings during this time. By now I was very interested to try to discover why he decided to attend Gresham College on that first meeting day. But I also puzzled as to just how a French spy came to know Oliver Cromwell's brother-in-law, let alone be invited to a meeting with so many disgruntled Parliamentarians, who unanimously elected Cromwell's brother-in-law to chair them!

Sir Paul Neile

Fifth on the list of founders is Sir Paul Neile. Neile was born in 1613 and had been a courtier to Charles I. For his service as an usher of the Privy Chamber he had been knighted in 1633. In 1640 he was elected MP for Ripon during the Short Parliament, but during Cromwell's rule Neile very wisely lived quietly, near Maidenhead, keeping a low profile. He remains almost invisible with little else recorded about him until the minute books of the Royal Society start to report some of his activities. It is clear that he was very much an amateur scientist whose particular skill was a patience in the grinding of optical glasses for use in telescopes. It was this private interest in the production of high quality optics which first brought together the then disgraced courtier and the powerful Warden of Wadham College. Indeed Neile had such skill at grinding lenses that John Wilkins preferred to spend his honeymoon with Sir Paul, just talking about the grinding process, rather than with his new bride. Perhaps this was a wise move, considering the age of Robina Cromwell.

In July 1660 the king reinstated Neile to his position as a Gentleman Usher of the Privy Chamber. This explains how Neile came to be in London on 28 November. Neile had used his exile in Maidenhead to make telescopes. He had also allowed Christopher Wren to use them to make observations of the planet Saturn. Once Neile was back in favour at the Court he became a go-between for the king and the Society, and he had the advantage of an existing friendship with the out of favour, but scientifically useful, Wilkins. The only other important item of information I could discover regarding Neile is that he was a founder of the Hudson Bay Company and that he had interests in merchant shipping.

I couldn't help but wonder who invited Neile along. It could have been Wren, since they were acquaintances, but Neile also lived in the Palace of Whitehall while attending the king – so he could hardly have avoided also knowing Sir Robert Moray. Or was it his old friend John Wilkins who encouraged him to come to the meeting?

Dr Jonathan Goddard

As I worked down the list of founders I came next to Dr Jonathan Goddard. Goddard was a medical man, who had obtained his doctorate of medicine from Cambridge in 1643, at the age of 26. He had been appointed Professor of Physic (an old term for medicine which gives us the modern word physician) at Gresham College in 1655, but he was, at that time, Warden of Merton College, Oxford. In other words Goddard had the best of both worlds. Was he allowed such licence because he was Oliver Cromwell's personal physician, I wondered? Whatever the reason he didn't move to Gresham until three years later, holding the Gresham appointment *in absentia.* He continued to live in Oxford, and to draw the warden's stipend, until Charles II summarily dismissed him. Goddard was friendly with both Wilkins and Ward while he was at Oxford. But when Charles purged Oxford of Parliamentarians, Goddard decided it was a good time to fall back on his Gresham professorship, and he moved to live in his College rooms.

Many of the early Society meetings were held in his rooms at Gresham. The College was very important when the Royal Society was being formed, and I decided I would need to try to find out more about it. Historian Gerald Weld, who carried out a full review of the early minute books of the Royal Society, had also noticed the important role of Gresham College and he said of it:

> *There is every reason to believe that the members of the College were very favourably disposed towards the infant Society of Philosophers.*[23]

I couldn't help wondering why so many Gresham professors supported a 'Royal' Society so soon after being thrown out of University posts by the newly restored king.

Dr William Petty

Dr William Petty invented modern statistics. He developed techniques of recording and analysing the detail of political events, involving large numbers of people, which laid the basis for the modern Office for National Statistics. Born in 1623 he gained his earliest education serving as ship's boy before joining the Royal Navy. He retained his interest in ships and shipping for the rest of his life. When the first Civil War broke out, Petty left England. He went to Paris to study medicine and chemistry and while he was there he met Thomas Hobbes and Descartes. He returned to London, after the defeat of the king, and was well placed when Parliament removed many of the incumbents of high office at the Universities and replaced them with their own supporters. Petty became a fellow of Brasenose College, Oxford and was awarded his MD. By 1650 he held the Chair of Anatomy at Brasenose and had also been created the Professor of Music at Gresham College. His real success, however, came when he took two years' leave of absence from his academic positions to go to Ireland as chief physician to Cromwell's army. This job he carried out extremely well but he now showed other talents. Once the Commonwealth army had subdued Ireland the seized lands had to be redistributed and new titles of ownership created. In December 1654 he offered to complete a new survey of the whole of Ireland within thirteen months. He succeeded brilliantly and his 'Down Survey' (so-called because Petty was based in County Down while he did the work) still forms the basis of legal record of title for a large proportion of the land holdings of Ireland.[24]

It was during his time in Ireland that Dr Petty met Robert Boyle, who became a patient of his and also his friend. Through Petty, Boyle met the 'Parliamentary High Table Group' (including Wilkins and Ward). These were academics who had replaced Royalists and now held all the senior positions at Oxford. Petty made a large amount of money from his successful survey of Ireland and became independently wealthy. However, he still held his Oxford and Gresham College appointments *'in absentia'* and still drew both stipends. In the late fifties Petty began to take a practical interest in the design of efficient sailing vessels. He started to work on designs for double-hulled (catamaran-type) vessels which had the potential to greatly outpace current ships.[25]

Petty was the first man to use statistical data in pursuit of political argument. He truly earned the title 'the father of modern statistics'. However, he had been such a strong supporter of Parliament, during the period of the Commonwealth, that in late 1660 he was stripped of the Vice-presidency of Brasenose College, Oxford. He went to live in London, keeping his head down with the other refugees. The only academic post he still maintained at that time was the Chair of Music at Gresham College. Perhaps it is hardly surprising that he met up with his old colleagues, who had also been ousted from their cosy University posts by the newly returned king. As he was in residence at Gresham College his attendance at Wren's lecture on 28 November 1660 did not surprise me, but why he wanted to help a Royal Society was a puzzle. He had no reason to like the king or hope for the monarch's patronage.

Mr William Ball

Mr William Ball, was an amateur scientist and a Royalist supporter. Just as Charles II had picked the first President of the Royal Society he also chose the first treasurer and his choice was William Ball. Prior to the 28 November meeting Ball had been cooperating with John Wallis to study the rings of the planet Saturn. I knew this because between 1656 and 1659 Wallis wrote a series of letters to the Dutch astronomer and mathematician, Christiaan Huygens. In these letters Wallis reported the results of Ball's observations to Huygens.[26] Huygens was later to quote Ball's work in his own theory of the nature of Saturn and its satellites entitled *Brevis Assertio Systematis Saturni*.[27]

Huygens visited Ball's London home on 1 May 1661. On the

evening of that visit Mr Ball held a dinner to celebrate the first anniversary of Parliament's reading of Charles II's Declaration of Breda.[28] The acceptance of this statement by Parliament had paved the way for the king's return from The Hague in May 1660. Sir Robert Moray, who had spent some years in the Netherlands, was also invited to the dinner.

William Ball was the Society's first Curator of Magnetics. Thomas Birch[29] wrote a great deal about the magnetic experiments carried out by Ball. He also noted that on 4 April 1666 the minute book says:

> *It was ordered that Mr Ball should be written to by Mr Oldenburg [the then secretary] to know what he had done in magnetical experiments, and that he should be desired withal to send up the magnetical apparatus, that was with him, belonging to the Society, who had present occasion for it.*[30]

Five months later Sir Robert Moray asked the Council's permission for Mr Ball to keep some of the equipment for his own use. The cost of this apparatus was eighteen pounds. At this time Mr Ball had moved to live on his father's Mamhead Estate, in Devon. Fortunately, the Council agreed to Sir Robert's request and so Ball was able to continue with his magnetic experiments. Soon after this Ball carried out the first trials of a method modern archaeologists call magnetostratisgraphy. This is a way of matching the alignment of naturally magnetised areas of the earth's surface with the present direction of the earth's magnetic field.[31] He had found an outcrop of loadstone in Devonshire and Robert Hooke, the Society's curator of experiments, suggested to him that he should 'observe how the poles lay in the earth, whether parallel to the axis or after the manner of the dipping needle or parallel to any meridian'.[32] On one occasion Ball created a magnetic needle ten feet in length to compare its accuracy with a standard mariner's compass.

So the main facts I could find about Mr Ball are that he was a Royalist who kept company with leading Parliamentarian academic John Wallis; he was an amateur scientist, interested in astronomy and magnetism; he was a close friend of Sir Robert Moray and he had impressed the king so much that his majesty insisted that Ball be made the first treasurer of the Royal Society. His presence at the meeting was hard to explain. Moray was the most likely candidate to have invited him but that just re-posed the question, why was Moray himself there?

Mr Laurence Rooke

Laurence Rooke was the host of the meeting of 28 November. At the time he was Professor of Geometry at Gresham College and aged thirty-eight years. He had gained his degree from King's College, Cambridge in 1643 and then retired for three years to live in the country. He seems never to have enjoyed good health. Indeed, he was not even fit for his own graduation. His degree was awarded '*in absentia*' as he was not strong enough to attend the ceremony. He went to live in Kent after completing his degree. This retirement to the country air seemed, however, to strengthen him and in 1650 he moved to Wadham College, to study under John Wilkins and Seth Ward. He also met, and worked with, Robert Boyle at Oxford. The fact that he was acceptable at Oxford suggests he must have been a Parliamentary supporter, as all Royalists had been ousted from the universities. After two years working there he was offered the professorship of Astronomy at Gresham College, a post he held for five years until he became the Gresham Professor of Geometry in 1657.

Rooke's main area of interest was in the measurement of longitude. He wanted to know how to find a ship's position in the open sea. His first ideas were to use sightings of the moon or the movements of the moons of Jupiter. He wrote papers on methods for observing lunar eclipses for 'the geographical purpose of determining terrestrial longitude'. Rooke knew that the movement of shadows on the moon's surface can be used as an accurate clock. The jagged peaks of the mountains of the moon act like the pointer on a sundial and he thought that the various craters and rifts could make up the scale of this celestial clock. As the moon was visible from everywhere on the earth's surface the moment of shadow contact happened at the same time for every watcher. Rooke recognised the moon as a giant sundial hanging high in full view of the whole world. All that was needed to know the longitude was to measure the altitude of a first magnitude star and compare it with its altitude at the same time for the homeport.

Charles II was so impressed with the idea that he asked for a demonstration showing this effect. His instructions, sent via Sir Robert Moray, asked for a large-scale globe model of the moon to be constructed 'representing not only the spots and various degrees of whiteness upon the surface, but the hills, eminencies and cavities moulded in solid work'.[33]

The model was built by Christopher Wren and presented to the king's private museum. It was set up on a rotating stand so that it could be illuminated and turned to reveal all the phases of the moon 'with the variety of appearances that happen from the shadows of the mountains and valleys'.[34]

The idea is ingenious and will work, if the sky is clear enough to allow a detailed view of the moon and the mariner is a skilled astronomer, familiar with the surface features of the moon. In addition, the sailor would need an ephemeris, a table showing the positions of the main stars.

This idea shows Rooke to have been an intensely practical man, very capable of original thought. This, however, did not extend to taking care of his own health. He caught a chill, while walking home without his coat, after a visit to the house of his patron the Marquis of Dorchester, and died on 26 June 1662.

The problem he was working on at the time of his premature death, the determination of longitude at sea, was the most important problem of the day. It was not completely solved until the invention of the marine chronometer, nearly one hundred years later, in 1759. Historian C A Ronan says of Rooke:

> *Rooke's work . . . became forgotten as the problem of longitude was solved. Moreover, excellent theoretician though Rooke may have been (and contemporary statements indicate that he was so thought to be) he did not make any major contributions either to astronomy or to mathematics and there is therefore no obvious reason for his inclusion in histories of astronomy or indeed in histories of science.*[35]

What a sad epitaph for a man who came so close to solving the navigation problem of ships at sea a hundred years before Harrison. Had he lived, the history of marine navigation might well have been very different and the loss of the HMS *Association* and her three sister ships that wrecked themselves on the Scillies in 1707 might have been avoided.[36] Although a founder of the Royal Society, Rooke never became a fellow as he died before the granting of the First Charter.

Sir Christopher Wren

But what of Christopher Wren, the man who made the scale model of the moon to demonstrate so graphically the usefulness of Rooke's big idea? Wren was a gifted model-maker, a skilled scientist and the best architect of his generation. Sir John Summerson said of him:

> *It seems the enigma of Wren's dual capacity as scientist and architect is not really a very profound one. A young man of exceptional gifts, with natural abilities as a draughtsman and model-maker, was drawn into a circle of men considerably older*

> *than himself. His remarkably elastic mind enabled him to come abreast with most of them in their own fields when, on nearly every occasion, his propensity for visual expression was made evident.*[37]

Christopher Wren was born on 20 October 1632 in a little village about sixteen miles from Salisbury. His mother died when he was only two years old and the following year his father, also called Christopher, was appointed Dean of Windsor and Registrar of the Order of the Garter. The earliest memories of young Christopher would have been those of living in the grounds of Windsor Castle and mixing with its Royal occupants. The Installation of a boy only slightly older than himself as the Prince of Wales and a Knight of the Garter must have impressed him. As Dean of Windsor, his father took part in the ceremony on 12 May 1638 and I couldn't help wondering if the six-year-old Christopher met the boy who would grow up to become Charles II or if he remembered the pomp and coldness of Charles I as he conferred these regal honours on his nine-year-old son.

Prince Charles Louis, the exiled Elector Palatine was also staying at the Deanery of Windsor. He was trying to persuade Charles I to support him in returning to his Electorate. The Elector had as his personal chaplain a young clergyman who has already figured in this story, John Wilkins. At this stage of Wren's life both he and the Revd Wilkins were clearly in the Royalist camp.

The event that seems to have decided Wilkins that he would fare better on the side of Parliament happened in 1642 as young Christopher was celebrating his tenth birthday. A troop of Roundhead soldiers, led by a Captain Fogg, seized the Deanery of Windsor and ransacked it. The Wren family fled first to Bristol and then to Bicester, near Oxford. (Wilkins fled to London. He did not side with the Royalists again until after the

28 November meeting and the Restoration forced his hand.)

Christopher Wren senior, however, remained a firm supporter of the king. First, at Bristol, and then later in 1645, after Bristol had fallen to Lord Fairfax, in Oxford. (Charles had moved his Court to Oxford at that time.) In an attempt to keep his son out of the hostilities Wren senior sent Christopher to school in London, where he studied under Dr Charles Scarborough.[38] While he was in London young Christopher again met up with John Wilkins, now a supporter of Parliament. Wilkins' reward for his support of the Parliamentary cause was the Wardenship of Wadham College, Oxford. In 1650, the eighteen-year-old Christopher went up to Wadham College to study under Wilkins. Wilkins became Wren's protector, something he certainly needed in those difficult times. In 1647 Wren's father had faced serious charges from Roundhead purists. They said that the decorative plaster work he had created in his Church at East Knoyle, where young Christopher had been born, was too ornate and papist! Wren senior was severely censured and lost his living while Wren junior prospered at Oxford, under the patronage of Wilkins.

Christopher Wren was developing rapidly both socially and intellectually by 1657, as by now he had become a Gresham Professor of Astronomy. While reading a miscellaneous collection of Wren family papers, published in 1750 by his grandson Stephen Wren, I discovered that Christopher had led a very exciting life at Oxford. And he met some interesting people!

He had become friendly with the family of Wilkins' niece by marriage and Oliver Cromwell's daughter, Elizabeth Claypole. She was three years older than Christopher and married to John Claypole, who had been Cromwell's Master of Horse in the battles against Charles II in 1651. Claypole served as an MP in both 1654 and 1656, which must have kept him in London a great deal. Elizabeth was also close to Oliver Cromwell. She was his second daughter, and reputedly his favourite.[39] One day, while Wren was at dinner with the Claypoles, Cromwell arrived unexpectedly and joined the dinner party. Cromwell spoke kindly to Wren and knew that Wren's uncle Matthew, the Bishop of Ely and a confirmed Royalist, was imprisoned in the Tower of London. Wren's grandson, Stephen Wren, recorded the following conversation between Cromwell and young Christopher:

> *'Your uncle has long been confined in the Tower.'*
>
> *'He has so, sir,' replied Wren, 'but he bears his Affliction with great Patience and Resignation.'*
>
> *'He may come out if he will,' was Cromwell's unexpected retort.*
>
> *'Will your Highness permit me to tell him this from your own Mouth?' Wren asked.*
>
> *'Yes, you may,' Cromwell replied.*[40]

Bishop Wren did not take up Cromwell's offer, as he was not prepared to swear allegiance to the Republic. He remained in the Tower until he was freed at the Restoration.

Stephen Wren also lists the interests and skills that Christopher developed while he was at Oxford. Among them are the following:

> *Hypothesis if the moon is solid; to find whether the earth moves; new ways of sailing; probable ways of making fresh water at sea; the best way of reckoning time-way-longitude and observing at sea; fabrick for a vessel of war; to build sea forts, moles, etc; inventions for better making and fortifying havens, for clearing sands and to sound at sea; to stay long under water; submarine navigation; easier ways of whale fishing; new cyphers; a compass to work in a coach or the hand of a rider; a new way of rowing.*[41]

This list shows a great interest in matters of naval warfare and navigation.

When Laurence Rooke was promoted to be Professor of Geometry at Gresham College in 1657, it left the Gresham Chair of Astronomy vacant. Christopher Wren was appointed to fill that vacant Chair. To mark his preferment Sir Paul Neile, an old friend of the Wren family from their days in Windsor at the court of Charles I, gave Christopher a new and bigger telescope.[42] Wren used this telescope to good effect during the four years he stayed at Gresham. Strangely, Wren left Gresham, in 1661, to take up the job of Savilian Professor of Astronomy at Oxford. This was the post Seth Ward had been ejected from by Charles II only twelve months earlier. Obviously by this time Charles had decided to forgive Wren's flirting with Cromwell and his family, but by then Elizabeth had been dead three years and Wren had lost touch with her husband. Perhaps Charles remembered his childhood playmate. If so, this put Wren in a position to repay the favours of protection and patronage that Wilkins had paid to him at Oxford.

It is not surprising that Wren was at the meeting on 28 November as he had just given the lecture that the others had listened to. It would have been only natural to invite him to the after-proceedings. What was puzzling about Wren was the fact that the other members wrote to him after the meeting to ask him to join them. Why did they need to do that if Wren had attended the meeting? Had he left before they got down to their real business, and if so, why?

Mr Abraham Hill

Finally I needed to consider the last person on the list, Mr Abraham Hill. He seems a very odd choice for a founder of the Royal Society. He was only twenty-five years old but early in 1660 both his parents died leaving him a moderate fortune. He had no need to work to keep himself and as he had not benefited from a university education he seems to have decided to take advantage of the public lectures offered by Gresham College, Oxford.

He was a regular listener to Wren's lectures and so it must have been natural for him to be invited to the discussions afterwards. He was certainly keen on the early experimental proceedings of the new Society, serving on many committees and assisting the more learned members with various experiments.[43] When William Ball retired as the Treasurer of the Royal Society in 1663, Abraham Hill replaced him. In March 1665, Sir Robert Moray made use of Hill's business skills, to make the formal application, on behalf of the Society, to Charles II for a patent concerning a new way of making watches and clocks for use at sea to determine longitude. As the procurement of a patent was expensive a number of other inventions were included. These were for several kinds of carriage, a powder horn, an apparatus for dressing hemp and for various types of guns and pistols.[44]

Hill was far more businessman than scientist. Sir Robert Moray seems to have recognised this and encouraged Hill to carry out many of the money-related activities of the Society, a job Hill was good at. He became very interested in the theory of money, and finance, and later went on to become comptroller to the Archbishop of Canterbury.

Conclusion

The twelve original founders of the Royal Society split into two major groupings. About half were Royalists who had kept out of public life during the rule of Cromwell and had returned to London to seek advancement at the court of King Charles II; and most of the other half were Parliamentarian academics who had taken control of the Universities under Cromwell but had been thrown out of virtually everywhere when Charles had returned, except Gresham College. Add into this mix one independently wealthy young man who was following a voluntary course in self-education, again at Gresham, and that is a pretty clear picture of the founders.

But how did two such wildly different groups ever come to be meeting socially and then just happen to decide to form a scientific society? I was hardly any nearer to answering that question. In fact if anything the mystery had deepened. Two younger members, who later became extremely important scientific fellows, seemed to have left before the rest of the group got around to discussing setting up a society and had to be asked in writing, afterwards, if they wanted to join. Had the older ten waited until the younger men had retired before daring to discuss their revolutionary ideas for natural philosophy?

What had become very clear was that only one of these original founders seemed to have any real influence with the king and that was Sir Robert Moray. But this ex-French spy, and monarchist rabble-rouser seemed out of place among the Parliamentary Puritans of the Gresham set. How had he come to be there at all?

Before I could hope to understand what had really been happening I needed to know more about the period around that first meeting. I knew that the minutes of the meeting said:

> *And to the end that they might be the better enabled to make a conjecture of how many the elected number of this Society should consist, therefore it was desired that a list might be taken of the names of such persons as were known to those present, whom they judged willing and fit to joyne with them in their designe, who, if they should desire it, might be admitted before any other.*[45]

Who did the ten decision-makers of these founding twelve consider to be fit and proper persons to work with them? This had to be the next question I set myself. Perhaps looking at the type of people they chose to join them would help me understand their motives.

CHAPTER 3

Conflicting Stories

A list containing the names of forty persons was therefore prepared, and to each of them was sent an invitation urging them to become members of the new Society in addition to the twelve who had already resolved to hold regular weekly meetings.[1]

Sir Henry Lyons FRS

THE LAST ACT OF THE first meeting of the Royal Society was to make up a list of people who were deemed suitable to become members of the newly formed group. The minutes do not say who proposed this idea but it is clear that it was taken up with enthusiasm. A list of forty names was drawn up. Sir Henry Lyons says of this choice:

> *The response to this appeal was very satisfactory, for of those whose names appear on the list only five did not become Fellows of the Society. Of the remaining thirty-five candidates nineteen may be considered as men of science while the other sixteen included statesmen, soldiers, antiquaries, administrators and one or two literary men.*[2]

Once again this seems very amiable and cosy, but just glancing down the list, Sir Henry presented, I saw the same strange mix of Royalists and Parliamentarians appearing. There was John Wilde, one of the stalwarts of the Rump Parliament who had just been thrown out his job in the Exchequer by Charles, and William Brereton, Commander-in-Chief of the Parliamentary Army in Cheshire. While on the Royalist side one of Charles II's new court favourites, Thomas Povey was there along with Elias Ashmole who, in appreciation of his service to Charles I as a master of Ordnance, had just been rewarded with the post of Windsor Herald.

A certain complacency of retrospective knowledge seemed to colour Sir Henry's account. He had done a good job of combing through the early minutes of the Society, but he continually assumes that the only possible motive of these strange political bedfellows was to form a scientific debating group. I still had doubts that this was a strong enough motive to bring such wildly opposing groups together in such evident harmony.

I decided to check out all the names on the list and to see what I could find out about the others with whom I was not familiar. What I discovered increased my doubts about their motives.

Of the forty proposed recruits, listed at that first meeting, ten had been consistently neutral in the dispute between king and parliament. I also recorded if the person proposed was an academic or had held, or recently been dismissed from, a political post. Of the politically neutral suggestions all were practising academics. Two had held Gresham Professorships, Daniel Whistler and William Croome. George Bate had been Court Physician initially to Charles I, then later to Cromwell and he then held the same post for Charles II. He could hardly have been more neutral! Francis Glisson was the Professor of Physic at Cambridge, George Smyth, George Ent and Nathaniel Henshaw were all practising medical men and members of the Royal College of Physicians. John Austin was a fellow of St John's College Cambridge and Thomas Willis was Sedleian Professor of Natural Philosophy at Oxford. The last of the men I categorised as politically neutral was Christopher Wren. He had been a childhood friend of Charles II and he had regularly taken dinner with Cromwell and his close family. If not neutral, he was at least flexible. I was still puzzled by the fact that even though Wren was listed as being present at that first meeting it had seemed appropriate to send him a formal invitation as well. I decided to put that puzzle aside while I considered the rest of the evidence. I next decided to check how many of the politically neutral academics had accepted the invitation to join. They all had. Only two Royalists and a Parliamentarian had turned down the invitation. (Sir Henry's other missing two, Dr Phrasier and Dr A Cowley, poet, turned out on closer inspection to be Sir Alexander Frazier and Abraham Cowley, Fellows number 142 and 61 respectively.) The Royalists were Henry Coventry, a courtier to Charles II while he was in exile in France and Thomas Rawlins, the chief engraver of coins at the Royal Mint. The Parliamentarian who turned down the honour was John Wilde, a longstanding member of the Rump Parliament. Evidently, when Charles ejected him from his post of chief baron of the exchequer he decided to take his bat home. I found only three of the men on that original list of forty had refused the invitation to join.

I had used an excellent monograph produced in 1982 by Michael Hunter for the British Society for the History of Science to check out the details of the first members. Hunter had produced a full catalogue of all the fellows (strange choice of title fellow, isn't it? I made a note to look into the origins of this title for elected members later) with details of who proposed whom, when they were elected, how active they were and whether or not they paid their subs. The very first fellow listed was the man whom Charles had insisted was to be President,

William Brouncker. The first eleven fellows listed were the men present at the first meeting with the man the king wanted as president as first fellow.[3] I decided that the order in which fellows were accepted into the Society was a good measure of how important whoever was running the Society considered them to be.

The second fellow to be made was Robert Boyle, who also appears on the list of forty persons to be invited to join ahead of all others. Again, I was puzzled as to why Boyle should need a separate invitation when he was listed as being at the first meeting, the gathering at which the list was drawn up! If he was there, why write to him? I made another note to return to the question of the need to write to Wren and Boyle at some future time and then looked at the data I was accumulating on the early members.

Boyle was an important catch for the fledgling society. He was the fourteenth child and the seventh son of an extremely fertile Earl of Cork and money had never been a problem for him. At eight years old he had been shipped across the Irish Sea to study at Eton; by eleven he was travelling around Europe in the company of a paid tutor; and at the ripe old age of fourteen he visited Italy to study, at first hand, the works of the recently deceased Galileo.

His private education kept him out of the academic hands of the 'Old Schoolmen', the peddlers of Aristotle's rigid view of the Universe, and his studies in Italy meant that in his formative years he was exposed to the observational science of Galileo rather than trained in the theoretical thinking of the Clerics who controlled both the Universities, and the Inquisition. His 'Grand Tour', however, was not entirely free from religious influence. While he was staying in Geneva he was caught outdoors in an intense thunderstorm, which so frightened him that he became extremely devout, gave up associating with women and refused from that time forward to ever take an oath.[4] Needless to say he died childless and a bachelor, but the time he saved in abstaining from wine, women and swearing was devoted to experimental science. In 1645 he became independently wealthy and able to afford equipment to further his interest in experiments. His father, who had rarely abstained from women and never from wine, died of his excesses and left the eighteen-year-old Robert a steady and sizeable income.

It was at this time Boyle had moved to Oxford and became an acquaintance of John Wilkins. Vacuums also started to fascinate him. By the age of thirty he had designed and built a machine to pump air. Boyle himself was never much of a technician but he employed, as a servant and handyman, a young man from the Isle of Wight, by the name of Robert Hooke. Boyle designed the pumps and Hooke built them and made them work. Boyle's early vacuum pumps were very temperamental affairs, in fact Hooke seemed to be the only person who could be trusted to make them function reliably. Nobody but Hooke, it seemed, could get the piston which moved the air to seal. However, with Hooke operating his air-pump, Boyle was able to succeed in carrying out an experiment which Galileo had only been able to dream about. Boyle placed a small lump of lead and a feather within a glass tube from which the air had been pumped. The lead and feather fell at exactly the same speed, through the vacuum. His fame was already assured when Wilkins and his friends, after seeing this experiment, took to calling a vacuum produced by a pump, 'a Boylean vacuum'.

Boyle also had the idea of placing one of the new-flanged ticking pendulum clocks inside the tube. While the container was full of air the clock could be heard clearly, but when Robert Hooke coaxed the temperamental pump to remove the air, the ticking could no longer be heard, although the movement of the pendulum showed that the clock had not stopped. All in all Boyle was a useful addition to the company, despite his total aversion to taking oaths, which would eventually prevent him becoming President of the Royal Society.

Of the original forty, twenty-four were academics, the remaining sixteen were all in influential political positions. In his *History* Sir Henry tries to show these politicians in a good light but to describe Sir John Denham as a poet is rather like describing Adolf Hitler as a painter. The statement is true but incomplete. Sir John had been a senior councillor to Charles I, he had escorted Queen Henrietta Maria during her flight to Paris, and had been trusted by Charles I to carry written instructions to Charles II after the late king's execution in 1649. Sir John had stayed in Holland until 1658, when he returned to England. He also wrote poetry, but his close links to the king would seem to be more important to explain why he was on the list of forty. In the same manner Sir Kenelm Digby was interested in chemistry, and was a courtier, but he had risen to fame as a successful naval commander for James I and had then spent many years as Chancellor to Henrietta Maria in France, before returning to England as Henrietta's ambassador for two years in 1654. He had gone back to Paris in 1656 and stayed there until the Restoration, after which he returned to London to stay.

Digby was another fortunate addition to the group as he brought a wealth of practical experience with him. Once the society was established he was quickly drafted into a subcommittee looking at matters of concern to the navy, an area where he had great experience, but this was not where Digby himself had expected to contribute to the knowledge of the society.

His main obsessions were his collection of plants and his interest in curative medicine. However, it must be said that his thoughts regarding medicine were less acceptable than his studies of plant behaviour. He proposed that all manner of wounds could be cured by the application of 'copperas' or 'green vitriol'. At first sight the idea has merit as the ferrous sulphate he described has both astringent and antiseptic properties, but Digby intended to apply the antiseptic, not to the wound but to the weapon that caused it! He called this 'cure' the powder of sympathy. He persisted in believing in this magical powder despite the fact it was rarely successful. On some occasions, though, he could be a careful observer, as his studies of the development of chicken embryos showed. These experiments were much more systematic and accurate, leading to an early insight into embryo development.

It was as a collector of plants, however, that Digby made his mark on science. He had noticed that in some circumstances his plants thrived while at other times they didn't. He collected a number of observations about when plants grew vigorously and when they did not. He then spotted that if he scattered Salt-Petre on the soil near his plants they flourished more. Perhaps this is how he decided that scattering ferrous sulphate around wounded patients might also help them re-grow new flesh. His skills of observation were, however, much better for plants than for people. He noted how a seed would become swollen with water and that this would make it sprout new growth, and likened this to the development of an embryo chick. He saw how plants in different atmospheres grew at different rates and came to the conclusion that the very air itself contained something important to plant growth. He tried some experiments growing plants within airtight bell jars and found that without a constant supply of fresh air the plants would not flourish. His early comments laid the foundations for our present understanding of how plants draw in and process nutrients and gases to grow.

Kenelm Digby published both his big ideas. His theory of wound treatment is now little more than an amusing footnote to medical history but his *Discourse concerning the vegetation of Plants* is honoured as the first scientific paper on horticulture.

As the Royal Society grew Digby's naval and navigational

knowledge would be put to practical use; his ideas about medicine would be ridiculed; but his observations on plant growth led to further experiments with Boyle's air pump, to prove the importance of air to a growing plant.

Elias Ashmole is remembered today as the Antiquary that Sir Henry describes but at the time he had just been made Windsor Herald, and had previously been a Royalist soldier. Just why Ashmole should be considered a scientist is at first sight a little strange. He was a lawyer and an historian. the sort of person that today would be considered a follower of the Arts. But he was also an astrologer. Today astrology is not considered to be a science; since Isaac Newton explained how the solar system really worked, astrology is thought to be little more than a superstition. However, this was not the case in 1660. Before Newton discovered the laws of astronomy, astrologers predicted the tides from the phases of the moon, as well as the fortunes of their clients. Astrology was a serious enough subject to be taught at the Universities. Early in the Civil War Ashmole stayed at Brasenose College, Oxford specifically to study the science of astrology. From March 1645 onward, Ashmole's diary starts to include the astrological calculations which he undertook every day.

Josten, Ashmole's biographer, says of astrology at this time:

> *Until Sir Isaac Newton promulgated the universal law of gravitation, astrology provided the only generally recognised universal law. Even the few scholars who repudiated 'judicial astrology', the branch of astrology which is concerned with prognostication, as an idle superstition, accepted astrological rules as a code linking the eternal and incorruptible celestial spheres to the corruptible sub-lunar world by an all-pervading system of sympathies and antipathies.*[5]

In the light of this prevailing attitude, which viewed astrology as the only true science, perhaps Kenelm Digby's strange views on powder of sympathy are more easily understood. It was considered a mark of learning to understand the rules of astrology, and the mathematical calculations needed to work out the rules of God's heavenly clockwork were as complex as any carried out by the proto-astronomers of the time. The whole science of logarithms and the first practical slide rule calculator were discovered by one of the leading astrologers of the time, William Oughtred, as part of his attempts to simplify the calculation of horoscopes.

Ashmole first learned how to cast a horoscope at Brasenose College but he went on to study the 'science' under Oughtred. Even the Church took a sympathetic view of astrology, and when William Lilly published a detailed textbook on Christian Astrology in 1647 its academic status was assured.

It was accepted wisdom that the positions of the celestial bodies at the moment of a child's birth impressed on an infant's soul the intentions of God. So the position of the stars could be used to foretell character, natural gifts, physical constitution and destiny. However, as Christian astrologer William Lilly was at pains to point out, the stars did not compel and, within the limits of a certain determination, man's will remained free. The purpose of astrology was to give a glimpse into the hidden mechanisms of life so that an individual's actions could be brought into harmony with the celestial influences.

Ashmole cast his daily horoscopes with the aid of a set of astronomical tables, known as an Ephemeris. The arithmetic was quite complex and Ashmole became skilled enough at the calculations, by the standards of his day, to be considered a mathematician. He cast horoscopes for himself; he cast horoscopes for clients, who paid him for his advice; and he consulted the stars on a daily basis to decide on his actions for the coming day. He predicted the most favourable moment to carry out particular tasks, and he cast horoscopes to answer specific questions by using the time the question crossed his mind to represent its moment of birth.

Nowadays astrology is considered at best a bit of fun and at worst a useless superstition but in 1660 astrologers were thought to be learned scholars. So Ashmole, as a respected astrologer, was already a 'man of science' at the time of his invitation.

Thomas Henshaw, whom Sir Henry describes as an historian, had served for many years under Sir Robert Moray in the elite Scots Guard of the king of France. He was brother to Nathaniel Henshaw who also figures on the list. Of the Political appointments, fourteen of the sixteen were Royalists, many of whom had a great deal of influence at the court of King Charles II. The two Parliamentary political nominees were John Wilde, the Rump MP who refused the invitation, and William Brereton who had been a senior officer in Cromwell's New Model Army before he became interested in natural philosophy.

The pattern seems to be quite clear. When the list was drawn up at the end of the first meeting, it consisted of a number of academics who were of neutral politics but who were well placed in the Universities and the Church; some disaffected but able Parliamentary academics who had fallen from favour after the Restoration; other Royalist academics who had been returned to their posts by the king; and a number of political and military heavyweights who now had influence at Court.

I couldn't help asking myself if this had been planned. The structure of the group suggests that having had the acquaintance of Sir Robert Moray in Charles II's court, in France or in Holland, helped you get on the list, particularly if you still had some influence at the newly restored English Court. There seemed to be a deliberate plan to create a fashion for supporting science among the wealthy courtiers of Charles II. It also seemed to be a deliberate intention to establish a system that would support some able academics who had fallen on hard times. In addition it appears to have been a concentrated attempt to attract as many as possible of Charles's personal physicians (who were also members of the Royal College of Physicians) to become Fellows.

While compiling the list I had noticed that links with two individuals kept cropping up. These were Sir Robert Moray and John Wilkins. I decided to run through the list again and mark those who I knew had either had contact with or could have had contact with Wilkins or Moray. For example I knew that Lord Hatton had been exiled in Paris for eight years during the period when Moray had been serving in Louis's Scots Guard, also in Paris. Using this technique showed a fascinating pattern of coincidences. For about eight of the forty I had no knowledge of possible links with either, but of the remaining thirty-two, fourteen had been in contact with Wilkins prior to that first meeting. Of these, thirteen were academics and only one was political, this individual being Matthew Wren, the Uncle of Wilkins' protégé Christopher Wren. All of the remaining eighteen had either been in contact with Sir Robert Moray or had the opportunity to do so. Of the eighteen, ten were influential figures in Charles's Court. Regarding the remaining eight academic appointees, five were court physicians to Charles and his family. At this time Moray was living in a grace and favour house within the grounds of the Palace of Whitehall.

None of the people who had been in contact with Wilkins or Moray failed to become fellows. It seemed prudent to review the order in which the Society turned the list of proposed members into Fellows. Using the Fellow number as a measure of priority it was clear that once the more able academics had been put in place the next priority was the politically influential. But it struck me that another factor was also at work. Those with an interest in naval matters all seemed to have lower fellowship numbers and so must have been considered more desirable as members.

The Well-prepared Go-between.

On 5 December 1660 the minutes of the Society show that:

> *Sir Robert Moray brought in word from the Court, that the King had been acquainted with the designe of the Meeting. And he did well approve of it, and would give encouragement to it.*

This was only a week after the very first meeting! Sir Henry, in his history, suggests that the academics were making use of Sir Robert Moray to further their scientific ends:

> *The way was now clear for those who had the matter in hand to press forward with their scheme for the formation of the Society which they had planned, and it was highly desirable that its aims and constitution should not arouse suspicions in the minds of the authorities. Here Sir Robert Moray was able to render invaluable service for he was not only well known to the king but was trusted by him; he was therefore a most suitable emissary to bring to the king's knowledge what the philosophers had done, and what they were proposing to do in organising their Society; and this he did without delay.*[6]

I was beginning to wonder just who had been making use of whom. Sir Robert was either extremely eager to please his new Parliamentary friends, or he had prepared his ground already.

During that second meeting the prospective members agreed to be bound by an obligation to support the Society and its aims. The wording was as follows:

> *Wee whose names are underwritten, doe consent and agree that wee will meet together weekeley (if not hindered by necessary occasions), to consult and debate concerning the promoting of experimental learning. And that each of use will allowe one*

> *shilling weekeley, towards the defraying of occasional charges. Provided that if any one or more of us shall thinke fitt at any time to withdrawe, he or they shall, after notice thereof given to the Company at a meeting, be freed from this obligation for the future.*

To the end of this binding obligation was appended the signatures of most of the attendees of the first meeting, the majority of the list of forty and a further seventy-three other names. This sudden burst of enthusiasm for membership led the fledgling Society to draw up conditions for membership.

On 12 December the following rules were passed, saying that 'no person shall be admitted into the Society without scrutiny, excepting only such as are of the degree of Barons or above'. They also agreed that 'Publick Professors of Mathematicks, Physick, and Natural Philosophy, of both Universitys, have the same privilege with the College of Physicians, they paying as others at their admission and contributing their weekly allowance and assistance when their occasions do permit them to be in London.'

I was interested to note that one of the first members to be scrutinised under this new rule and to be found acceptable was Sir Kenelm Digby, the Admiral turned Courtier.

By March 1661 the young Society had set up a governing committee system which elected a chairman each month. From that time until the granting of the First Charter, which made Brouncker the permanent President, a number of individuals served in the office. The first chairman under this rule was Sir Robert Moray. This man served this office for nine months during the Society's gestation period. Wilkins served five times and Boyle and Brouncker once each. Had Moray used Wilkins to reassure the disaffected Roundheads, before taking on a more visible role later in the process? While regularly chairing meetings Moray, as his correspondence with Huygens shows,[7] was also devoting considerable time to the committee which was drafting the proposed Royal Charter.

It was difficult, however, to find any link between Wilkins and Moray. Sir Henry's history takes the line that I had already found difficult to reconcile with the facts. He suggested there were two separate groups of Natural Philosophers, one each side of the Civil War, and that they decided to form the Royal Society to further their scientific aims. Under this scenario Sir Robert Moray was made use of by Wilkins to found a scientific society. But from the very first meeting it seems to be Moray who was making the running. Wilkins is down on his luck, forced out of his University post, earning scraps by itinerant preaching and scrounging a bed from Seth Ward. He must have been too busy trying to rebuild his life to have had the free time to spend instigating a very expensive philosophical society. No matter how I looked at the facts Sir Robert Moray had to be the driving force, but what could possibly link him to Wilkins and the newly disenfranchised academics? And how did the common interest in naval matters figure in this scenario? Moray's links with the Royalists, freshly returned from exile, were clear; I needed to know more about the group Wilkins had been involved with and there were two major clues to investigate: a series of letters written by the young Robert Boyle about an 'Invisible College', and an account of the early pre-Royal Society meetings given by John Wallis and later withdrawn.

Looking at the Invisible College.

The first of Robert Boyle's comments on what he called the Invisible College was in a series of letters written to his former tutor Isaac Marcombe in Paris, during October 1646.[8] One comment in particular caught my eye:

> *The best of't is that the cornerstones of the Invisible (or as they term themselves the Philosophical) college, do now and then*

honour me with their company, which makes me sorry for those pressing occasions that urge my departure.[9]

Boyle was eighteen years old when he wrote this letter. The name Invisible College was his own term for a group of men who met regularly to discuss what he described as:

natural philosophy, the mechanics and husbandry according to the principles of the philosophical college, that values no knowledge but as it hath a tendency to use.[10]

Boyle came across these men in London between 1646 and 1647. In a letter to Francis Tallents, of Magdalene College, Cambridge, he wrote more about these 'cornerstones of the Invisible College':

Men of so capacious and searching spirits, that school-philosophy is but the lowest region of their knowledge; and yet though humble and teachable a genius, as they disdain not to be directed to the meanest, so he can but plead reason for his opinion; persons that endeavour to put narrow-mindedness out of countenance, by the practice of so extensive a charity, that it reaches unto every thing called man, and no other less than an universal goodwill can content it. And indeed they are so apprehensive of the want of good employment, that they take the whole body of mankind for their care.[11]

These references struck a particular resonance with me. I was aware that other writers had speculated that the Royal Society might have been the organisation which gave birth to Freemasonry. The term 'cornerstone' is used by Freemasons in a special and important way during the ceremony of admitting a new member into a Lodge. Part way through the ritual (known as the Initiation) the new member, known at the Candidate, is placed in the Northeast corner of the lodge and then this speech is made to him:

> *Brother . . . It is customary at the erection of all stately and superb edifices to lay the first or foundation stone at the Northeast corner of the building. You, being newly initiated into Masonry, are placed in the Northeast part of the Lodge, figuratively to represent that cornerstone, and from the foundation laid this evening may you rise a superstructure, perfect in all parts and honourable to the builder.*

So any man who has become a Freemason has gone through a ritual where he had acted out the part of a cornerstone. Freemasons can recognise and identify one another by using quotes from the ritual. All the rituals of Freemasonry have very fixed verbal forms, often involving question and answer which have to be totally memorised word perfectly. Any Freemason can pick up a question in the ritual and expect another Freemason to give the correct answers. Often one Freemason will recognise another by his 'Masonspeak', the odd phrases he introduces into normal speech. In these letters Robert Boyle is using Masonic forms to describe the members of what he has termed the Invisible College. His mention of the equality of opinion and of the scope of charity all have Masonic resonances. But first let me, for the benefit of non-Masonic readers, explain the purpose of the Second Degree of the Freemasonry, normally called the Fellow-Craft degree. When the Fellow-Craft Freemason is awarded his special apron, which carries a distinguishing embroidered badge, a speech is made to him telling him the purpose of the degree:

> *Brother . . . I must state that the badge with which you have been invested points out to you that, as a Fellow of the Craft, you are expected to make the liberal arts and sciences your future*

study, that you might better be enabled to discharge your duty as Mason and estimate aright the wonderful works of the Almighty Creator.

The new Fellow Craft is then moved to another part of the lodge and this speech is made to him:

You now stand to external appearances a just and upright Fellow Craft Freemason . . . in the former degree you had the opportunity of making yourself acquainted with the principles of moral truth and virtue, you are now in this Degree permitted to extend your researches into the more hidden paths of nature and science.

Finally as the new Fellow of the Craft completes the ritual of the Second Degree he is given a further item of information:

As a Fellow Craft, you may, in our private assemblies offer your sentiments and opinions on such subjects as are regularly introduced in the Lectures, under the superintendence of an experienced Master. By this privilege you may improve your intellectual powers, qualify yourself to be a useful member of society and, like a skilful brother, strive to excel in the good and the great.

During the Initiation ritual the Candidate is asked to donate what he will towards Charity. The following speech is then given to impress on him the importance of Charity:

I shall proceed to put your principles in some measure to the test, by calling on you to exercise that virtue which may justly be denominated the distinguishing feature of a Freemason's heart – I mean charity. I trust that I need not here dilate on its excellence; doubtless it has often been felt and practised by you;

> *suffice it to say it has the approbation of heaven and earth.*

All Boyle's references to his Invisible College echo these Masonic sentiments. In a letter to Samuel Hartlib, Boyle again uses a Masonic analogy of building a living temple of the intellect:

> *And since you do not disdain the meanest workman that is but willing to lay some few stones towards the building of your college.*[12]

It is unlikely that Boyle himself was ever a Freemason. He refused to become President of the Royal Society as it involved taking an oath, which he would not do. He could not have become a Freemason without taking an oath, so if he is using Masonic terms and ideas he must have acquired them from somebody else. Was Boyle's Invisible College really an early lodge of Freemasons?

The first complete history of the Royal Society was published in 1667 and it was written by Thomas Spratt under the direction of Brouncker, Moray, Wilkins and Evelyn. As historian Margery Purver pointed out it can therefore be considered a definitive view of what the founders of the Royal Society wanted to record about their origin: She said of Spratt's *History*:

> *The History is the only publication that ever received from the Royal Society such supervision in its documentation; and this scrutiny was carried out by those who were chosen for their personal knowledge of the facts. It shows that Spratt was not speaking for himself nor for any other private person, but for the Royal Society as an institution, which considered this book to be its special concern, the first comprehensive and public account of its origin, policy and business.*[13]

Spratt says that the origin of the Royal Society was a series of meetings held at the lodgings of John Wilkins in Oxford between 1648 and 1659[14] but John Wallis, who became the Savilian Professor of Geometry at Oxford in 1649, wrote a letter which made different claims.

Wallis was a mathematician who rose to fame by studying very small intervals of time. He laid the foundations for a new type of maths called calculus, which Isaac Newton developed in order to analyse the orbits of the planets. Wallis was interested in the problem of what happens when you need to carry out calculations involving dividing one number by another, when both of the numbers are getting very small. The smallest possible number is zero, but there is no real solution to the problem of dividing of zero by zero so Wallis asked the question, how close to zero can I go before the calculations become meaningless?

Wallis never quite managed to answer that question, that honour went to Newton, but Wallis did a lot of the basic thinking which inspired Newton. When Newton said he 'stood on the shoulders of giants', at least one of the giant's shoulders belonged to John Wallis. Newton said of Wallis: 'About the beginning of my mathematical studies, as soon as the works of our celebrated countryman, Dr Wallis, fell into my hands, by considering the Series, by Intercalculation of which, he exhibits the Area of the Circle and Hyperbola, he inspired me to open up the integral.'[15] This inspired 'opening up of the integral' led Newton to discover the basic principles underlying today's rocket science.

Wallis, in his turn, had been inspired to study the problem of the arithmetic of tiny time intervals when he met the astrologer William Oughtred, inventor of the slide rule. Wallis said his first textbook on mathematics was Oughtred's book, *Clavis Mathemiticae*. In this book Oughtred developed ideas for methods of predicting the positions of planets in the sky. He wanted to improve his astrological predictions, but really had no mathematical techniques he could use to predict small orbital movements. He knew these movements were happening because of the changes in the azimuth and declination of the planets. These he could see from his naked eye observations. But he needed something to help him carry out the complicated maths needed to work out exactly what was happening. The calculating machine he invented was the slide rule.

Wallis was an excellent mathematician with an inherent skill for spotting patterns. This served him in good stead during the Civil War, when he worked as a code-breaker for Cromwell, and he was rewarded with the Savilian Professorship at Oxford.

It seems incredulous that Wallis, a scientist, author of the first book on Algebra,[16] the thinker who inspired Newton to solve the problem of predicting the clockwork of the Universe, the man who laid down the basic rules of algebraic notation – which still plague today's school children – learned his scientific skills from an astrologer! It shows just how far science has developed in today's world, when these two topics, astrology and physics, once considered aspects of the same subject are now thought to be as different as Richard Dawkins and Mystic Meg. Wallis, however, had mixed with astrologers, alchemists and academic theologians in his younger days and in a pamphlet written in 1678, six years after the death of John Wilkins and five years after the death of Sir Robert Moray, he wrote:

> *About the year 1645, while I lived in London (at a time when, by our civil wars, academical studies were much interrupted in both our Universities), beside the conversation of divers eminent divines as to matters theological, I had the opportunity of being acquainted with divers worthy persons, inquisitive into natural philosophy, and other parts of human learning; particularly into what hath been called New Philosophy or Experimental Philosophy. We did by agreements,*

divers of us, meet weekly in London on a certain day and hour, under a certain penalty, and a weekly contribution for the charge of experiments, with certain rules agree amongst us to treat and discourse of such affairs; of which number were Dr John Wilkins (afterwards Bishop of Chester) then chaplain to the Prince Elector Palatine in London. Dr Jonathan Goddard, Dr George Ent, Dr Glisson, Dr Merret (Drs in Physick), Mr Samuel Foster. then Professor of Astronomy at Gresham College, Mr Theodore Haak (a German of the Palatinate and then resident in London, who I think gave the first occasion and first suggested those meetings) and many others.

These meetings we held sometimes at Dr Goddard's lodgings in Wood Street (or some other convenient place near), on occasion of his keeping an operator in his house for grinding glasses for telescopes and microscopes; sometimes at a convenient place (The Bulls Head) in Cheapside, and (in term time) at Gresham College at Mr Foster's lectures (then the Astronomer Professor there) and, after the lecture ended repaired, sometimes to Mr Foster's lodgings, sometimes to some other place not far distant.

Our business was (precluding matters of theology and state affairs) to discourse and consider of Philosophical Enquiries . . . About the year 1648/9 some of our company being removed to Oxford (first Dr Wilkins on his appointment by the Protector as Warden of Wadham College, then I and soon after Dr Goddard) our company divided. Those in London continued to meet there as before (and we with them, when we had occasion to be there) and those of us at Oxford, with Dr Ward (since Bishop of Salisbury) Dr Ralph Bathurst (now President of Trinity College in Oxford) Dr Petty (since Sir William Petty), Dr Willis (then an eminent physician in Oxford) and divers others, continued such meetings in Oxford and brought these studies into fashion there.[17]

What is remarkable about this account is the passing mention of rules which governed the conduct of the meetings he describes. Wallis clearly states that the topics of religion and politics were forbidden at the meetings. The words 'under a certain penalty' is the same wording used on the summons to many modern Freemasonic lodges. There was only one organisation in existence at that time which forbade the discussion of religion and politics at its meetings; meetings held to discuss the hidden mysteries of nature and science, and that organisation was Freemasonry.

Sir Henry Lyon's view is that:

> *The philosophers very wisely were content to follow the lines on which they had worked for several years until such times as a more formal organisation could be safely introduced.*[18]

But the whole scenario Wallis describes, of regular meetings, of the group from London splitting up and forming new groups elsewhere, fits the organisation of Freemasonry at the time. At least two of the twelve founder members at the first meeting were recorded as Freemasons, Bro Sir Robert Moray and Bro Alexander Bruce were both members of the Lodge of Edinburgh.[19] Also one of the men on the list of forty was a well-known Freemason, Bro Elias Ashmole. Was Freemasonry the link which explained how Sir Robert Moray came into contact with the Wilkins set?

Conclusion

The additional list of proposed members, which was drawn up by the founders at their second meeting, showed very similar characteristics to the founding group. It drew from both sides of the Civil War and almost everybody on the list took up the invitation to join. A large majority of the list were known to either John Wilkins or Robert Moray. None of the people they knew had turned down the invitation to join.

The first official History of the Royal Society had been written by Thomas Spratt, under the supervision of Wilkins and Moray. It made the claim that the idea had been hatched at a series of meetings hosted by Wilkins, during his time at Oxford. Two of the early Fellows, Robert Boyle and John Wallis, however, made different claims. They both wrote about other formative groups and each used symbols and ideas which are characteristic of Freemasonry to describe these other meetings.

If Wallis and Boyle are to be believed there were groups meeting under Masonic conditions and using Masonic symbolism for their discussions.

All the Wilkins' set had connections with Gresham College and its professors, so I decided I needed to find out more about Sir Thomas Gresham and his endowment. However, the traditional view has always been that the Wilkins group were following the teaching of Francis Bacon. Did Bacon have any links to Freemasonry?

It seemed evident that I would have to investigate further the possibility of a Masonic connection between Wilkins and Moray. I was aware that I could not yet propose a motive for Moray's actions but perhaps a more detailed look at what had been happening to Freemasonry under the Stuart kings might help provide me with an answer.

But first I needed to know more about the importance of Francis Bacon in the events leading to the formation of the Royal Society. Did he have any connection with the Freemasonry of the time?

CHAPTER 4

The Patron Saint of Frozen Chickens

[Bacon] was remarkably blind to the important scientific work that was going on in his own time . . . he ignored the brilliant work of his own doctor, William Harvey, on the circulation of the blood . . . dismissed Gilbert's theory of magnetism, as a kind of occultist fantasy . . . disdained Copernicus and ignored Kepler and Galileo. Nevertheless, there is no question about the degree of respect in which he was held by British scientists of the succeeding generation.[1] Anthony Quinton

JOHN WILKINS PRESENTED the Society with the very first copy of Thomas Spratt's newly finished official *A History of the Royal Society*, at one of its meetings. This presentation copy had an engraving by John Evelyn on the frontispiece. Under this plate Wilkins wrote: 'Presented to the R. Society from the Author by the hands of Dr John Wilkins, Octob. 10 1667.'

Evelyn's engraving demonstrates the importance the founders of the Royal Society attached to Francis Bacon. The plate shows a room high above a distant landscape. The rear of the room has an open bay window beneath a dome. Through the two leftmost windows there is a rural landscape, while beyond the right window is a distant view of Gresham College. The front aspect of the bay is supported by an arch and two pillars. Hanging from the keystone of the arch is the coat of arms that

Charles II awarded to the Society.

The room itself has a black and white squared floor. On the left-hand side is a bookcase, containing a library of knowledge. On the ledge of the bookshelf stands a diploma with a Royal Sceptre on top of it. There are many tools hanging around the walls of the room: four sets of compasses, three squares (devices for testing a right angle), and two plumb-lines. In addition there is a telescope, a long case clock, a small marine clock (mounted within a keystone), two globes and various pestles and mortars.

A bust of Charles II stands on a single pillar, directly under the keystone of the arch and in the centre of the black and white pavement. Above the head of this bust a winged angel holds a laurel crown of fame. To the left of Charles sits William Brouncker, who was then President of the Society. He is pointing to the inscription on the pillar, which announces Charles to be the Royal Founder and Patron of the Society. To the right sits Francis Bacon, with a collar around his neck supporting a suspended jewel formed from a pair of crossed keys. In his left hand he holds a purse, with the Royal Crest on it and with his right hand he points towards the tools hanging from the pillar behind him. Under his portrait is an inscription describing him as the inspirational source of the Society.

When I first saw this plate I was amazed at the use of so much symbolism, which, had I seen it included in a more modern engraving, I would have said was Masonic. Is it purely coincidence that John Evelyn made so much use of symbolism that is still used in present day Freemasonry?

I decided to interpret what I, as a Freemason, could see in the picture. The first impression is the Masonic pavement in the forefront of the picture. It pushes towards the viewer so that it cannot be ignored. All Masons are told about the black and white chequered floor of the lodge room: 'The Mosaic pavement is the beautiful flooring of the lodge, by reason of its being variegated and chequered, This points to us the diversity of objects which decorate and adorn creation, the animate as well as the inanimate parts thereof.'

The compasses and squares, of which there are four compasses and three squares in the plate, are described in Masonic ritual as follows: 'The compasses and square, when united regulate our lives and our actions. The compasses belong to the Grand Master in particular and the square to the whole craft.' The four sets of compasses are obvious while the squares are well in the background.

The sceptre (or its secular form, the mace) lying on the library shelf, on top of a diploma, says two things, knowledge is power and it echoes the statement made in the First Degree that 'In every age Monarchs themselves have been promoters of Masonry, and have not thought it beneath their dignity to exchange the sceptre for the Trowel.'

The room has three pillars, two supporting the arch and one supporting the bust of Charles. The two rear pillars when conjoined are said to represent stability. The ritual says, 'For God hath said in strength will I establish my house that it will stand fast forever.' At the keystone of this arch of stability, locking it firmly together, stands the coat of arms of the new Royal Society. The pedestal which supports the bust of Charles is positioned where the pedestal of Enoch would be placed in a Masonic Royal Arch Chapter. Masonic ritual says the pedestal is one of two pillars made by the Patriarch Enoch, who carved on them the secrets of all the sciences when the world was threatened by a great flood. Freemasonry says that the pillar was found in a secret vault when the site was being cleared ready for the building of Solomon's Temple. By placing Charles on the pedestal of Enoch, he is being portrayed as the saviour of science and civilisation against the forces of chaos.

Finally there is the positioning of the three figures. The seating of the officers of a lodge of Freemasons is very carefully controlled. Charles is placed as the Grand Master in the East, with the light of the rising sun behind him. Brouncker is placed in the seat of the senior working officer, while Bacon is placed in the seat of the immediate Past Master.

Of course I realise that all this Masonic symbolism could be coincidental, but with at least three well-known Freemasons among the first members of the Society, it could also have been quite deliberate.

There is also some non-Masonic symbolism. Brouncker is pointing at Charles as the Patron and deferring to him. Bacon is holding a tightly closed purse with a Royal Seal on it and is pointing into the darkness to the right-hand side of the plate. And in the background the building of Gresham College can be seen, indicating where the Society has originated.

Spratt's *History* makes some points about what the Society intended to achieve which seem to echo the symbolism of placing Charles on the pedestal of Enoch:

> *If now this Enterprise shall chance to fail . . . the world will not only be frustrated of their present expectation, but will have just ground to despair of future labours . . . This will be the last great endeavour that will be made in this way, if this shall prove ineffectual; and so shall not only be guilty of our own ignorance but of the errors of all those who come after us.*[2]

Perhaps the symbolism of the Society as the keystone in the Arch of Stability, and the king as saviour of the arts and sciences of civilisation was not accidental, as the same sentiment is echoed within the body of the book. Is the other Masonic symbolism just as deliberate?

But, as Anthony Quinton has already pointed out in the opening quote of this chapter, Bacon was not the best scientist or philosopher of his generation. Why was he held in such high esteem by the founders of the Royal Society? I needed to know more about the man if I was to answer this question.

The Vision of a New Atlantis

Francis Bacon was born on 22 January 1561 at York House, just off the Strand in London. His father Sir Nicholas Bacon was a friend of William Cecil, who at the time of Francis's birth was an envoy to Scotland for Elizabeth I. Cecil would become Elizabeth's chief minister, with the title of Lord Burghley. Francis's father became Lord Keeper to Elizabeth. So, from an early age Bacon knew the senior members of the English Establishment.

He graduated from Trinity College, Cambridge, in 1576, at the remarkably young age of fifteen years three months. His biographer Dr William Rawley, said of him at this time: 'He fell into the dislike of the philosophy of Aristotle; not for the worthlessness of the author, to whom he would ever ascribe all high attributes, but for the unfruitfulness of the way.'[3] Dr Rawley, who was chaplain and amanuensis to Bacon in his declining years, may be attributing an early insight to him which evidence from Bacon's life does not support.

At the age of sixteen Francis Bacon was admitted to Gray's Inn, in London. By the age of twenty-one he was a qualified barrister. Two years later he was MP for Melcombe Regis and had started upon a political career; first under Elizabeth I, who does not seem to have taken to him; and then under James VI(I) under whom he prospered. He was knighted within the first four months of James's rule but did not achieve any high office until 1607, when he became Solicitor-General. He had already published his book *The Advancement of Learning* in 1605 and by 1610 had drafted *The New Atlantis*, which would not be published until after his death. He was forty-four years of age before he appears to have had any ideas at all about science, but from then on he wrote profusely. Was he a late developer? Or did he only come to ideas on how to study science after James became king of England?

Thomas Spratt had no doubts when he wrote about the importance of Bacon's ideas:

> *It must be first of all begun, on a scrupulous, and severe examination of particulars; from them, there may be some general Rules with great caution drawn: But it must not rest there, nor is that the most difficult part of its course: It must advance those Principles, to the finding out of new effects, through all the varieties of Matter: and so both courses must proceed orderly together; from experimenting, to demonstrating, from demonstrating to experimenting again.*[4]

Abraham Cowley, one of the men named on a list of forty-first fellows of the Royal Society contributed a poem to Spratt's *History* that likened Bacon to Moses:

Bacon, like Moses, led us forth at last,
The barren Wilderness he past,
Did on the very Border stand
Of the blest promis'd Land,
And from the Mountains Top of Exalted Wit,
Saw it himself, and shew'd us it.

The choice of Moses as a simile is again interesting because in the Masonic Ritual of the Holy Royal Arch, whose symbolism appears to have been used in the frontispiece, the Royal Arch Mason, called a Companion, is told that Moses was one of the First Three Grand Masters of Freemasonry who held the First or Holy Lodge on the summit of Mount Horeb.

Bacon's career prospered under James. He became, in turn, Lord Keeper and then Lord Chancellor. By 1618 he was Lord Verulam and three years later became Viscount of St Albans, but then in 1621 everything went wrong. His past actions caught up with him. He was accused of corruption and admitted the charges. This trial, and its outcome, was a personal disaster for him. Being found guilty he was sentenced to a term of imprisonment in the Tower, subject to the king's pleasure. He was excluded from Court, disqualified from Parliament and fined £40,000. King James released him from the Tower after only three days, but Bacon's public life was over. He had to sell up his London house to pay the fine. Bacon retired to less expensive housing and spent his last five years writing. He completed his monograph on Henry VII, *Historia Ventorum*, in 1622 and *Historic et Mortis* the following year. In 1623 he published *De Augmentis*, a considerably enlarged version of a book, *The Advancement of Learning*, which he had first written eighteen years earlier.

He had completed *Novum Organum*, just before his disaster, and seeking to reconcile himself with James, he sent the king a copy of the book. James responded by likening it to a well-known biblical allusion. 'Like the peace of God, it passes all understanding,' he quipped.[5]

Bacon demonstrated his spectacular lack of practical experimental skill by dying in the cause of science while inflicting unspeakable indignities on a dead hen! In doing so he has become immortalised as the unofficial Patron Saint of Frozen Chickens!

John Evelyn tells the story in his diary:

> *He [Bacon] was taking the air in a coach with Dr Witherborne (A Scotsman, Physician to the king) towards Highgate, snow lay on the ground, and it came into my lord's thoughts, why flesh might not be preserved in snow, as in salt. They were resolved they would try the experiment presently. They alighted out of the coach, and went into a poor woman's house at the bottom of Highgate Hill, and bought a hen, and made the woman exenterate it, and then stuffed the body with snow, and*

> *my lord did help do it himself. The snow so chilled him, that he immediately fell so extremely ill, that he could not then return to his lodgings in Gray's Inn, but went to the Earl of Arundell's house in Highgate, where they put him into a good bed warmed with a pan, but it was a damp bed that had not been lain in about a year before, which gave him such a cold that in two or three days he died of suffocation.*[6]

At the age of sixty-five Francis Bacon died, a martyr to the cause of frozen food. Take a moment to reflect on his sacrifice the next time you visit the supermarket and see the plastic wrapped products of his final, fatal experiment!

His last work, which he had begun writing in the early stages of James's rule, was now published posthumously. It was titled *The New Atlantis*. This book began a debate about the nature of science that has continued down the centuries to the present day. But what did he say that was so revolutionary?

Bacon was the first writer on science to propose that a discovery is only scientific if it is guided by facts and not misguided by theory. In other words he proposed that when studying the hidden mysteries of nature and science the natural philosopher should both observe and experiment before proposing a theory. Bacon said that a scientific mind is a *tabula rasa*, a blank page devoid of all content, so that it can receive the imprint of nature without distortion.

The New Atlantis is interesting because in it Bacon sets out the ideas of a research establishment and how to divide labour within the study of science. The book, which was published a year after Bacon's death by William Rawley, is an adventure story about a ship lost in the South Seas. The adventurers happen upon a lost island, called Bensalem, where they are cautiously welcomed ashore. The inhabitants of Bensalem have the very first knowledge-based economy. At the centre of their civilisation is Solomon's House or the College of Six

Days' Works. The members of this house have chosen to live on this remote island in order to hide from the world's contagion.

All the personnel employed in Solomon's House have particular tasks which combine together to form a vast study of science. Some members extract material from books and others carry out experiments, while a proportion collate the results of these experiments. Still more members travel, while others work on technical applications or devise new experiments. The masters of Solomon's House turn all the cooperative labours into organised coherent theories.

The final section of the book deals with a vision of the future in which Bacon envisages a system of rituals that focus the minds of its followers upon science and technology.

In his biography of Bacon, Anthony Quinton said about *The New Atlantis*:

> *It is generally agreed that the idea of Solomon's House was at work in the minds of those who founded the Royal Society.*

Bacon himself said of it:

> *The purpose of Solomon's House is the knowledge of causes, and secret motions of things; and the enlarging of the bounds of human empire, to the effecting of all things possible.*[7]

Bacon seems to have crystallised ideas that were starting to appear around the beginning of the seventeenth century and formed them into a system of thinking about natural philosophy. The circumstances of his death show he was not a very practical experimenter, being much more concerned with the ideas behind the study of natural philosophy. He had developed the ideas in his second edition of *The Advancement of Learning* over the previous twenty years. But was it just coincidence that his first serious attempt to develop notions on how science should be approached occurred in the second year of the reign of James VI(I)? Bacon had not addressed the question of how to study the mysteries of nature before James arrived from Scotland. Did he learn of some new philosophy from the court of James VI(I)?

In 1609 Bacon wrote a rational analysis of the truths that may be hidden in the myths and fables of antiquity. He titled this book *The Wisdom of the Ancients*, but it was in *Novum Organum* that he published the ideas that would influence the founders of the Royal Society.

The frontispiece to the copy of *Novum Organum* that Bacon sent to James VI(I) also had a very Masonic-looking symbol engraved in it. It shows a ship sailing between two free-standing pillars. Margery Purver interprets the engraving as:

> *showing ships sailing through the Pillars of Hercules, the symbolic limit of classical science.*[8]

But to a Freemason it has a different interpretation. A new Mason is told how the two pillars which stood outside Solomon's Temple are a symbol of the strength and stability of a Fellow of the Craft.

If Bacon was a Freemason the symbolism of this frontispiece would advertise that the writer is a Fellow of the Craft and James, as I will show, was a Freemason. Bacon was deep in disgrace when he sent this book to the king. Perhaps the hidden message of the frontispiece of Bacon's book was intended to plead with the king for mercy. For the same Fellow Craft ritual says elsewhere:

> *You are not to palliate or aggravate the offenses of your brethren but to judge with candour, admonish with friendship and judge with mercy.*

Perhaps the message worked, because Bro James VI(I) allowed Bacon his freedom after only a token period of imprisonment.

Spratt, under the close supervision of Freemason Sir Robert Moray, had defined the objective of the Royal Society as being:

> *to overcome the mysteries of all the works of Nature for the benefit of human life.*[9]

He goes on to explain:

> *And this is the highest pitch of human reason; to follow all the links of this chain, till all secrets are open to our minds; and their works advanced, or imitated by our hands towards the settling of an universal, constant and impartial survey of the whole Creation.*[10]

Was it possible that Francis Bacon had been using the ideas and symbolism of Freemasonry when he created the frontispiece, and content, of the book he presented to the king?

Conclusion

Francis Bacon had never been a particularly good scientist but in the later third of his life he took an interest in techniques for studying nature. It was, however, only after the arrival in England of King James VI(I) that this interest in science developed.

Bacon is depicted on the frontispiece of Thomas Spratt's *History*, which had been supervised and endorsed by Wilkins and Moray, amidst a welter of Masonic symbolism. Bacon also made great use of Masonic symbolism in his own writings and on the covers of his books. Again, his use of Masonic symbolism only began after the arrival of King James in London.

I decided that I needed to look more closely at the history of King James VI(I) and how he came to become king of England. I already knew that James was connected to Scottish Freemasonry but did he have any connections with science?

CHAPTER 5

An Alien Monarch

'Freemasonry still persists and is ever on the forward and upward march. It attracts ever men with high ideals, humane ideas and wide vision . . . In Scotland the veritable proceeding of the lodges for the year 1599, as entered in their minute books are still extant. In England there are no Lodge minutes ranging back even into the seventeenth century.' Dudley Wright, Editor, Gould's History of Freemasonry

THE FAME OF JAMES VI of Scotland, the first king of Britain, really starts in 1603 with the death of Queen Elizabeth. The line of the House of Tudor died with her and her English crown passed to the son of her executed rival, Mary Queen of Scots. Young James had been acclaimed king of Scots in 1567, at the age of one, when his mother abdicated. Thirty-six years later Mary's son travelled south to London, there to be crowned James I of England.

The Tudor monarchs had been such strong rulers that they had made the task of governing England seem simple, but they had taught the English a respect for law. James, who was absolute monarch of the alien country of Scotland, was wholly ignorant of English ways and customs. During the final stages of his journey to Westminster, as he travelled through London, a thief was caught picking the pockets of the crowd who had assembled to meet him. James condemned the man without trial and sentenced him to be hanged out of hand. This type of disregard for the rule of law had not been a feature of Tudor rule and it was an early indicator that James was altogether unskilled in judging the English political climate.[1]

He was certainly a well-educated and clever man, able to read Latin and quote theology. He believed himself to be a divinely appointed king. To question the absolute authority of his 'royal prerogative' was, in James's eye, a treasonable insult to his crown and an impious blasphemy. This was not an attitude which endeared itself to the majority of the Protestant subjects of his new country. Indeed the first problem of James's reign was one of religion.

The various Protestant sects of early seventeenth-century England had only one common cause. They felt that the English Reformation had not gone far enough to really rid the land of Popery. They had approved of Henry VIII abolishing the Catholic Church and making the monarch the supreme governor of the Church of England but now each wanted the king to favour its particular rituals above all others. Most of the members of James's parliament were Protestants and pushed their anti-Papal views but the king had no intention of changing the settlement Elizabeth had imposed on the Church of England. He had suffered enough at the hands of the opinionated Kirk of Scotland and wasn't going to promote any other Church ruler but himself.

James saw an opportunity to deprive the Churchmen of their chief propaganda weapon by commissioning a definitive version of the bible. The choice of bible translation was being used to great effect by the more extreme Puritans. They used a version of the Bible with annotations which supported their extreme views. These annotations had taken on the power of scripture and James knew how to remove this advantage. In June 1604 he had instructed Bishop Bancroft of London to put together a team of translators and James himself sent them this message via the good bishop:

> *His Majesty, being acquainted with the choice of all them to be employed in the translating of the Bible, doth greatly approve of the said choice. And for as much as his Highness is very anxious that so religious a work should admit of no delay, he has commanded me to signify to you in his name that his pleasure is, you should with all possible speed meet together in your University and begin the same.*[2]

James laid down fifteen rules of translation that left the learned editors no scope for bigoted footnotes. In this way he intended to prevent the various extreme sects choosing translations which furthered their own political ends. He certainly managed to motivate the translators because the work, carried out at three sites, Cambridge, Oxford and Westminster by fifty academics, was completed in record time and by 1611 it was in print. James had managed to standardise the bible quotes his critics could legitimately use.

His motives are still clear from the preface the translators dedicated to him in the first edition of the King James Bible:

> *So that on one side we shall not be traduced by Popish persons at home or abroad, who therefore will malign verses, we are poor instruments to make God's holy Truth to be yet more and more known unto the people, whom they desire to keep in darkness; or if on the other side, we shall be maligned by self-conceited brethren, who runne their own ways and give liking to nothing but what is framed by themselves.*

As supreme governor of the Church of England, James had taken control of God's Truth and now his bible would give him a monopoly on its written form. Unfortunately, his lack of concern regarding other people's religious views pleased nobody. In 1605 the Catholics tried to blow up him, and his Parliament, with a great many barrels of gunpowder. However, the 5 November plot failed when Guy Fawkes was caught, almost in the act of setting the fuse, in the cellars of the palace of Westminster.

James never got on really well with his Parliaments, even though amid the emotion brought about by the Gunpowder plot he said what an honour it would have been to die in the company of his faithful Commons. He was extremely annoyed when his faithful parliament reminded him that they had traditional liberties including free speech, free elections and freedom from arrest during parliamentary sessions. James brushed these concerns aside but only through Parliament could he legally tax people, and when he found himself short of money, he had to deal.

His eternal problem was trying to reconcile his 'Divine Right' to rule with a Parliament that controlled his money supply. Ancient custom said that only Parliament could grant the right to levy tax. James soon discovered that his financial position was being undermined by a flood of precious metals from the New World. This was causing price inflation in Europe but the revenues from his Crown lands remained the same. He was getting steadily poorer and was forced to keep returning to Parliament for more money. He began to develop a theory of Government based on the need for an absolute monarchy and he found himself a brilliant advocate in the person of Francis Bacon. Bacon had risen to high office by supporting James's authority against the opposition of the Judges. Then Bacon fell foul of Parliament and the judges took their revenge by impeaching him for taking bribes.

James's opinion of Parliament was not very high, as the following comments he made to the Spanish Ambassador show:

The House of Commons is a body without a head. The Members give their opinions in a disorderly manner. At their meetings nothing is heard but cries, shouts and confusion. I am surprised that my ancestors should ever have permitted such an institution to have come into existence. I am a stranger, and found it here when I arrived, so that I am obliged to put up with what I cannot get rid of.[3]

For the whole of his 22-year reign James maintained an uneasy truce with his parliaments, never really approving of them but managing to avoid serious confrontations by avoiding calling them together, unless in dire need of money. In the early years of his reign his limitations as an English king, when compared to the example set by the Tudors, had been mitigated by the hope that Prince Henry of Wales would eventually succeed him. Henry was a bluff and hearty man who showed a dignified manner in his dealings with state matters. It appeared to Parliament he had the makings of a good monarch in the best Tudor tradition. They saw in Henry a future king who would not ignore their hard won ancient privileges or force the Royal Prerogative down their throats, as his father continuously did. When Henry died in 1612 this hope was extinguished and Charles, the new Prince of Wales was neither as statesmanlike as Henry, nor able to command the same respect his dead brother had.

James's court became characterised by his interest in attractive young men who became his favourites and could do no wrong. The most notorious of these were the Earl of Somerset and the Duke of Buckingham. James, however, had a further interest in male bonding activities.

The Freemason King

On the wall of the Lodge of Scoon and Perth hangs a painting of a very important Masonic event; the initiation of King James

VI of Scotland. The official entry of the lodge on the Roll of the Grand Lodge of Scotland simply says the lodge existed before 1658. This date refers to the charter of the lodge and is a set of rules which explain how the lodge was governed. The document which is signed by the Right Worshipful Master, J Roch, and two Wardens, Mr Measone and Mr Norie. This same charter records the event depicted on the wall of the lodge room. The charter states:

> *In the reigne of his Majesty King James the sixt, of blessed Memorie, who, by the said John Mylne was by the king's own desire entered Freeman, meason and Fellow-Craft. During his lifetime he mantayned the same as ane member of the Lodge of Scoon, so that this lodge is the most famous lodge within the kingdom.*[4]

The Mylne family figure a great deal in the early history of Freemasonry, no less than three generations of them[5] held the Mastership of the Lodge of Scoon and Perth between the late sixteenth century and 1658, when the Scoon Charter says the Mastership passed to James Roch. This same James Roch was the signatory to the document which records the making of James VI as a Freemason in 1601.

Another John Mylne, son of the John who initiated King James, had carved a statue of the king in Edinburgh in 1616. In 1631 this particular John Mylne was appointed Master Mason to Charles I and in 1636 resigned the office in favour of his eldest son, also named John Mylne[6] who had been made a Fellow Craft of the Lodge of Edinburgh in 1633. The third John Mylne took part in the Masonic meeting in Newcastle in 1641 where Sir Robert Moray was made a Mason. So the grandson of the man who initiated James VI initiated Sir Robert Moray.[7] There is every reason to believe that Sir Robert Moray received a family account of the initiation of James VI and could have been well aware of the close links between the Stuart kings and Freemasonry before he became personally involved with them.

But the real question I needed to ask was how did James VI, King of Scots come to be made a member of a lodge of Freemasons? To answer this question I had to review the documented origins of Freemasonry and its links with the kings of Scotland.

Early Scottish Freemasonry

The earliest documentary evidence for the existence of Freemasonic rituals is to be found on the south wall of a small, fifteenth-century Church in Mid-Lothian, now known as Rosslyn Chapel. It was built between 1441 and 1486 by William St Clair who was Earl of Caithness, third and last St Clair Earl of Orkney; Baron of Roslin; and Lord Sinclair. As Chancellor and High Admiral of Scotland, he was the second most powerful man in the kingdom. Indeed, he seems to have threatened the power of the Stuart kings. At the time when the earlier James II of Scotland was becoming deeply embroiled in English politics, William began to build what was then known as Roslin chapel. When James II's involvement in the Wars of the Roses lost him his life, at the Battle of Roxburgh, his son James III of Scotland stripped William of the earldom of Orkney and forced him to split his land between his many children. So the St Clairs' power was broken and they were never again strong enough to challenge the Stuart's grip on the crown of Scotland.

The new chapel William was building at Roslin was a tremendously ambitious project. All of the surface of the building, inside and out, was to be carved with tremendously ornate detail. Father Hey, the historian of Roslin, tells us that William personally supervised all the decoration, insisting that every piece was first carved in wood and presented for his inspection before he signed it off, marking the wooden test piece with a pass mark which allowed it to be committed to stone.[8] If Father Hey is correct then William St Clair was the first exponent of Quality Control, in his building works.[9] The implication of this statement is that none of the strange tableau carved into the structure are there either by accident or by whim of individual masons with a sense of humour. The fox, wearing clergyman's robes, standing in a pulpit lecturing to a congregation of chickens, tells us something about Sir William's opinions of the priests of the Church. However, it is a small tableaux on the external south-western corner which is the earliest documentary evidence of what is today known as speculative Freemasonry.

The scene shows a man kneeling in a very strange posture: his feet are placed in the form of a square, in his left hand he holds a bible, he is blindfolded and has a running noose about his neck. Alongside him stands a bearded man, robed as a Knight Templar, holding the noose. This strange pair are placed between two pillars. Except for the medieval clothing of the kneeling man, this scene could be a depiction of a modern Masonic First Degree Ceremony. Once I had realised how many points of similarity there are between this carving and a modern Freemasonic First Degree, I commented on this many times in Masonic lectures I gave. Eventually I published the evidence in a book co-authored with Christopher Knight.[10] On many occasions I was asked to debate this evidence, within Masonic lodges and on American and British radio. During these debates the Librarian of the Grand Lodge in London suggested that my conclusions that this was evidence of Masonic ritual in use in Scotland in the mid-fifteenth century could be explained away as simple coincidence. It just so happens that one of the subjects I teach at my own university is Statistics. In one of my regular lectures I look at the wider scope of statistical analysis in helping to understand evidence, so, as a demonstration, I decided to undertake a careful analysis of the suggestion that the similarities with William St Clair's authorised carving and the modern First Degree of Freemasonry were pure chance. My results were conclusive. Even if I gave the highest possible probabilities to things happening by chance there is only a probability of two parts in a thousand that all the similar elements to the modern First Degree are there by pure accident. To a statistician the facts say that William did not mix all these disparate elements by accident, unless he was incredibly lucky. In other words, he probably intended to have all those factors together when he approved the piece. The same 'landmarks' survive into modern Freemasonry, which claims to have preserved them from 'our antient brethren'.

This piece of evidence disproves the hypothesis that the elements of Freemasonic rituals included in the tableaux could have been there by pure chance, as the Librarian of the United Grand Lodge of England, who is not a statistician, had suggested. It leaves intact the alternative hypothesis I had put forward, namely that the ceremony was known to the builder of Roslin Chapel in the mid-fifteenth century.

The next piece of evidence linking Masons to a ritual connected with Solomon's Temple was noticed by historian Professor David Stevenson of St Andrews University. The evidence comes from Aberdeen. In the west front of King's College, Aberdeen, is a Latin inscription which Stevenson translates as:

> *By the grace of the most serene, illustrious and ever-victorious King James IV: On the fourth before the nones of April in the year one-thousand five-hundred the Masons began to build this excellent college.*[11]

Professor Stevenson goes on to point out that the date is significant for Freemasons, as it is the date traditionally accepted as that on which the building of Solomon's Temple started.[12] Stevenson comments further about the importance of Freemasonry in Scotland saying:

> *This, however, does not explain the peculiar wording of the inscription. It mentions the king as patron of the project but states that 2 April was the date on which the masons started work. It is surprising that an inscription of this sort should specifically mention the craftsmen responsible for the work at all and yet here they are standing alongside the king.*[13]

He adds:

> *By the late sixteenth century the Craft was in fact on the verge of a remarkable development which would make it different . . . one man saw that some aspects of the traditional heritage of the craft of masonry linked up a whole series of trends in the thought and culture of the age, and worked to introduce them in the Craft.*[14]

That man was William Schaw, who on 21 December 1583 became Master of Works to King James VI and Queen Anne.[15] He is buried in Dunfermline Abbey and that is where I went to find out more about him.

William Schaw, The Great Architect of the Craft

The old man who was my guide wore a dark suit, a white shirt and black tie. He took my hand and I felt the distinctive grip of a Fellow of the Craft as he led me towards Dunfermline Abbey. I smelt just a trace of mustiness as we entered the building. After the warmth of the morning sun, the air felt chilly and slightly damp, as we crossed the main nave through the shadows of the west wall. The old Mason pointed upwards towards the curving void of the roof.

'Those lads knew how to build square and true,' he said. His voice reverberating from stones our antient brethren had carved so carefully.

He was right! The great sweeps of the roof are supported on two central rows of towering, carefully carved pillars. The building had a commanding presence which made me want to lower my voice.

'Where exactly is Schaw's tomb?' I asked in a voice which now felt right as a whisper.

My guide took me towards the dimly lit north-west corner of the abbey. As my eyes adjusted from the bright sunlight to the striated light from the high window of the great church I could just make out an ornate monument against the north wall.

'The Auld Earl of Dunfermline built that for Schaw. He was a brother, initiated into the Lodge of Aberdeen, y'know,' the old Mason told me. 'He had a mark just like a lightning strike.' His walking stick scraped and clattered as he sketched the shape of the Earl's Mason's Mark on the floor as I watched. 'They all used their marks to identify themselves in those days, because no many o'them cud read o'rite.'

We stood side by side looking up at the tomb of William Schaw. There was a long inscription in Latin, telling of Schaw's life and works but my guide pointed to a smaller square inscription, high on the face of the monument, set between two pillars. He translated it for me, his voice strong as a bible reading, rolling round the quiet inner expanse of the Abbey:

> *Live in Heaven and live for ever, thou best of men. To thee this life was toil, death was deep repose. In honour of his true hearted friend William Schaw. Alexander Seton, Earl of Dunfermline.*

As the last echoes of his voice died away I took a closer look at the stonework of the tomb. The most striking thing about the monument, even more noticeable than its obvious expensive magnificence, was the number of mason's marks carved into it. It was as if every mason who had played any part in its creation had wanted to become a permanent part of it. The man who had taken me to see it was a senior brother from Dunfermline Lodge. I had been speaking to the lodge the evening before and had mentioned in passing the importance of Schaw in the creation of the lodge system. Afterwards this elderly gentlemen had come up to me.

'Have ye ever seen Schaw's tomb?' he asked.

'No,' I'd replied.

'If ye meet with me outside the Abbey tomorrow morning I'll see ye right,' he said. And he had.

I reached out and touched the cool surface of the stonework, tracing with my fingertip the outline of a mason's mark in the form of a five-pointed star.

'Ye know that mark, do ye? he asked.

'I know it was the mark Sir Robert Moray took at Newcastle,' I said.

'That's right and it's been used by many a good mason since,' my friend told me. 'But I like to think that Sir Robert might have marked his respects to Schaw.' I agreed with him. The five-pointed mark probably hadn't been carved by Sir Robert, but I felt spiritually very close to him as I traced the shape of his mark in cool stone work of the tomb of the man who invented the modern lodge system of Freemasonry.

William Schaw was born around 1550 in Clackmannan, near Stirling. His father John Schaw of Broich had been keeper of the king's wine cellar. By the age of ten William was employed at court as a page to Mary of Guise; I knew this because the Queen Dowager's accounts record his name on the list of her retainers for whom mourning was purchased. That same year his father John was charged with murdering the servant of another Laird.[16] William next appears in Scottish records when he signed the Negative Confession, a document which James VI and his courtiers had to agree to in order to assure the Reformed Church that the king and his retinue were not trying to bring back the Catholic Faith. William was a Catholic but seems to have been flexible enough in his religious attitudes to stay out of trouble with the Kirk. Professor Stevenson says of him:

> *Like a number of other Scots in court circles, though remaining a Catholic he avoided actions that might provoke persecution, probably attending Protestant services from time to time.*[17]

It was towards the end of 1583 that Schaw became James VI's Master of Works. About a month after his appointment the king sent him on a diplomatic mission to France, suggesting he had diplomatic as well as building skills.[18] This seems to be confirmed by the fact that the king chose Schaw to help entertain the ambassadors of the king of Denmark, who came to Scotland trying to negotiate the restoration of Orkney and Shetland to Denmark.[19] Schaw must have got on well with the Danes because in 1589 James sent him back to Denmark to escort his new bride, Anne of Denmark, to Scotland. Schaw went on to become Queen Anne's Chamberlain and a great favourite of hers. As his monument recalls:

> *Queen Anne ordered a monument to be set up to the memory of a most admirable and most upright man lest the recollection of his high character, which deserves to be honoured for all time, should fade as his body crumbles into dust.*

It was in 1590 that Schaw began to take an interest in Masons and their organisation. This first written evidence is a letter written under the authority of the king's Privy Seal to Patrick

Copland of Urdoch (near Aberdeen) confirming his right to act as 'wardanie over the maister masons of Aberdene, Banff and Kincarne'.[20] Professor Stevenson believes that Schaw may have been considering reorganising the mason Craft under a number of regional Wardens and used the historical precedent of the Coplands of Urdoch to re-establish the principle of regional Wardens. This information certainly shows that early Scottish Masonry was speculative and involved the higher ranks of society from a very early period. Patrick Copland was Laird of Urdoch, not a common workman.

However, eight years after confirming the authority of a regional warden in Aberdeenshire by privy seal, Schaw took on himself the role of General Warden of the Craft of Scotland. The post was a new one which Schaw created and it had the approval of a number of unnamed 'maister maissounis' who attended a meeting on the Feast of St John in Edinburgh in 1598.

Schaw, as the king's Master of Works, acted as agent for the throne in all state building works. This gave him a great deal of control over the Masons of Scotland and so he was only rationalising a state of affairs that was already in existence. His first Statutes contained 22 clauses.

The first clause insists that all Masons 'observe and keep all the good ordinances set down before, concerning the privileges of their Craft set down by their predecessors of good memory and that they be true to one another and live charitably together as becomes sworn brethren and companions of the Craft'. Here Schaw is referring to a system of regulations still known to Masons as the Antient Charges.

The remainder of the clauses deal with how the lodges shall be ruled and governed and how the work of the masons should be managed. There are two particularly interesting items. One seems to be the first health and safety directive ever issued to the building trade. It says:

> *That all masiteris [Masters], in charge of works, be very careful to see their scaffolds and ladders are surely set and placed, to the effect that through their negligence and sloth no hurt or harm come unto any persons that works at the said work, under pain of discharging of them hereafter from working as masiteris having charge of any work, but they shall be subject all the rest of their days to work under or with another principle masiter having charge of the work.*[21]

Stern discipline indeed for any Master Mason who did not take care that his workers were properly secured when they worked in the dizzy heights of a great cathedral or a Scottish grand house. Today's factory inspectorate would not quarrel with the intentions and sanctions of this sixteenth century Masonic legislation.

The other interesting item concerns how the Master of a lodge shall be chosen:

> *That there be a Warden chosen and elected each year to have the charge over every lodge . . . to the effect that the General Warden may send such directions to that elected Warden as required.*[22]

The Master of the Lodge [Warden] has to be elected each year and Schaw, as General Warden of the Craft, intends to issue any instructions via the elected officers of the Lodge to the Masons. This is a highly democratic system being put in place fifty years before the Civil War was to address the same questions of democratic accountability in England.

All in all, this was a far-sighted and fair document which has the obvious intention of simplifying the general management of Masons in Scotland. It takes account of the antient traditions of the order and respects existing rituals; it makes proper provision for safe working practices and it provides for regular democratic feedback from the *Maisteris* of the lodge. It was issued with the endorsement of all the Master Masons who had attended the Feast of St John meeting in Edinburgh in 1598 as the closing sentences show:

> *And for fulfilling and observing of these ordinances, set down as said here, the group of maisteris here assembled this day binds and obliges themselves hereto to be faithful. And therefore has requested the said General Warden to sign them with his own hand, to the effect that an authentic copy hereof may be sent to every particular lodge within this realm.*[23]

This document also represents the first time that any lodge had been instructed to keep written records of its proceedings.[24] The oldest lodge minutes in existence are those of Edinburgh, which start immediately after this meeting with Schaw.

The First Schaw Statute says a lot about Freemasonry. It confirms that Freemasons meet in lodges, that these lodges are ruled by Masters or Wardens, that there was a system of meetings at a higher level than the lodge, that lodges are obliged to keep written records of their activities and that they are honour-bound to observe the antient ordinances of their Craft. All of these components have survived down to modern Freemasonry and this is the earliest written evidence of their introduction. In other words, Schaw formalised the present day system of Masonic lodges. A lodge is not just the building where Masons meet; it is also the body of men who make up that group. It has its own traditions, hierarchy and records to prove what is has decided but is basically a democratic unit, inherited from a period when democracy was not supposed to have been prevalent.

However, there is more to this story because a well-established lodge existed out on the west coast of Scotland. This lodge, known today as Mother Kilwinning, was not based in Edinburgh but on the coast of Ayr, in the gr[illegible] Kilwinning Abbey. The Wardens of Mother Kilwinning Lodge were accustomed to issuing charters to other groups of masons so that they could form themselves into new lodges. What is more they claimed rights over the Mason Craft in Ayrshire. Schaw's First Statute, did not recognise the place that Kilwinning claimed in the newly created Masonic ranking. The following year, 1599, on the Feast of St John, Schaw issued his Second Statute, this time from Holyrood House, one of the king's palaces. This Second Statute accepted the statements in the First Statute but went on to assign a formal status to Kilwinning Lodge.

When Schaw had held his first formal meeting as General Warden of the Craft, again on the Feast Day of St John the Evangelist, Mother Kilwinning Lodge sent Bro Archebald Barclay to present a case that they should have a role in the new way of ruling the Craft. Bro Barclay was successful in making his case because Schaw now confirmed Kilwinning would be allowed to keep its antient practice of electing its officers on the eve of the winter solstice. It was assigned the rank of 'second lurge [lodge] of Scotland' and its Wardens were to have the right to be present at the election of all other Wardens of lodges within Lanarkshire, Glasgow, Ayr and Carrick. A Warden of Kilwinning was to have the power to summon and judge all Wardens of lodges within this area, with power delegated by Schaw as General Warden of the Craft. The Wardens of Kilwinning were to conduct regular tests of Masons within their jurisdiction to ensure they were properly trained in 'the art and craft of science and of the antient art of memory'.

With this clarification of the most important of the antient ordinances, and the adjustment to the pecking order between the Lodges of Edinburgh-St Mary's Chapel and Mother Kilwinning, Schaw seemed to have settled Freemasonry into a stable structure. Despite this he had greater ambitions for his fledgling new organisation. Schaw wanted the king to become Grand Master of the Order and he sought a Royal Charter confirming this status on the Craft for ever. He had one problem. The Masons would not accept a non-mason[25] as their Grand Master. Even though he was king, if James was to become Grand Master Mason, he would first have to be made a Mason.

In 1584 William Schaw had assisted his close friend Alexander Seton[26] in designing a house for Lord Somerville. The master mason employed to carry out the work was John Mylne.[27] In 1601 Mylne was Warden of the Lodge of Scoon and Perth. This lodge was situated in Scoon, which is the ancient place of coronation of the King of Scots. Here was an appropriate lodge for the king to join Freemasonry.

I have already quoted the minute of this event but now its political purpose had become clear. To complete his designs for the Craft, Schaw needed the king to be a Mason. James VI loved ritual, masques and dressing up. From all accounts he will have delighted in the ceremony at which he was initiated into the antient mysteries of the Mason Word. Schaw now had everything he needed in place to propose a Royal Grand Master Mason for the Craft; to be followed with the issue of the Royal Charter to confirm his authority as Lord General Warden of the Craft. Unfortunately the Masons of Scotland had different ideas. They claimed a different Grand Master, William Sinclair, Laird of Roslin.

William the Wastrel

When Schaw had promulgated his Second Statutes he had been on the verge of obtaining Royal sanction for the privileges of the Craft. Then he seems to have been forced into backtracking. A powerful group of Masons insisted he issued a document now known as the First St Clair Charter. Stevenson says of it:

> *[It] can be seen as indicating . . . that Schaw was forced to change his plans [for obtaining a Royal Charter] to take account of claims of the Craft . . . which were too strong for him to resist . . . Schaw's death in 1602 and the move of the king to England on the union of the Crowns the following year may have disrupted attempts to win the king's support.*[28]

After he had confirmed the claims of Kilwinning in its role as a minor Grand Lodge, the other lodges had recognised that Schaw could be put under pressure and might be coerced into modifying his opinions. Obtaining the agreement of the Lodge of Scoon and Perth to initiate king James VI had been Schaw's first move towards uniting the Lodges of Scotland under the Grand Mastership of the king.[29] The consequence of James's making the Lodge of Scoon and Perth his mother lodge would be, as the minutes say, 'so that this lodge is the most famous lodge within the kingdom'. This move would undermine all the jockeying for position which had gone on earlier. Edinburgh had already been named as first lodge, Kilwinning was officially number two and Stirling positioned third in seniority, but now, as the Royal Grand Master's Lodge, Scoon and Perth was poised to take precedence over all other lodges. By initiating the king, the Lodge of Scoon and Perth was outflanking all their brother Masons.

Schaw was now put under pressure by the lodges in the East of Scotland, Edinburgh, St Andrews, Haddington, Aitchison Haven and Dunfermline to acknowledge another antient authority in Freemasonry, that of William St Clair of Roslin. Stevenson comments about him:

> *Though in William Sinclair the masons had found a gentleman of ancient lineage willing to be their patron, they had not found a respectable or influential one . . . If the masons had had a free choice in seeking a suitable patron to advance the craft's*

interests they would never have chosen the laird of Roslin![30]

William St Clair, third and last St Clair Earl of Orkney, and the builder of Rosslyn Chapel[31] had been the second most powerful man in Scotland until 1471, at which time he had been forced by the king to split up his holdings. The Baronies of Roslin and Pentland had then been transferred to one of his son's Oliver, Lord Sinclair. Via him they had passed first to another William and then to an Edward before vesting in the particular William Sinclair, who is the subject of this charter.

This specific William Sinclair was a Catholic, and a man who kept falling foul of the Kirk. He used Rosslyn Chapel to have one of his children baptised in 1589. Rosslyn was not a Parish church but William was unperturbed by the outcry this caused. The minister who conducted the service, however, was forced to make a public plea for forgiveness.[32] A year later the presbytery of Dalkeith accused Sinclair of 'keiping images and uther monuments of idolatrie' in Rosslyn. The Kirk officials had to postpone interviewing him, however, as he had been arrested and charged with threatening the king's person.[33] When he was freed the Kirk pursued him, insisting that Rosslyn should not be used as a place of worship and that William force his tenants to attend the parish Kirk. They also suggested he set an example and become an elder of the Kirk. William declined saying he was 'insufficient' for the position. He proved his point soon afterwards when he was forced to make a public confession of fornication with a local barmaid. To add insult to injury, he told the Kirk he could not remember if all of his bastards had been baptised. When he was ordered to do public penance for his acts of fornication, by sitting on the repentance stool, he refused – unless he was supplied with a quart of good wine to help him pass the time.[34]

From the number of summonses to keep the peace and to refrain from attacking various individuals he seems to have been fond of both wenching and brawling. Father Hey, the historian of the Sinclair family, described him as 'a lewd man, who kept a miller's daughter for the purpose of fornication'.[35] He eventually ran away to Ireland with his mistress, abandoning his wife, son William,[36] and the Craft of Scotland.

This then was the man whom the Masons of Scotland preferred as their Patron, rather than allow the Lodge of Scoon and Perth take precedence over them. William the Wastrel, as the Laird of Roslin was known at the time, had the authority of the last St Clair Earl of Orkney behind his claim. Beyond this he was the keeper of the most important Freemasonic shrine in Scotland, Rosslyn Chapel. The Freemasons of Scotland let loose the only shot they could to thwart the ambitions of the Mylne family. The claim of the Laird of Roslin could be supported by appealing to the first sentence of the First Schaw Statute, 'that they observe and keep all the good ordinances set down before concerning the privileges of their Craft by their predecessors of good memory'.

The First St Clair Charter takes just this line when it says:

> *Be it known to all men that the Deacons, Maistres and Freeman of the Masons with the realm of Scotland with the express consent and assent of William Schaw, Maister of Work to our Sovereign Lord do assert that from age to age it has been observed amongst us that the Lairds of Roslin has ever been Patrons and Protectors of us and our privileges like as our predecessors has obeyed and acknowledged them as Patrons and Protectors.*[37]

So it would seem that Schaw's attempt to obtain a Royal Charter for the Freemasons failed because some lodges insisted on adhering to an older tradition which linked them to the Sinclairs of Roslin. The outrageous character of the man to whom they gave their loyalty suggests that the tradition must have been important to them, otherwise they could have gone along with Schaw's plan and taken Bro His Majesty King James VI, as their new Royal Patron. Certainly the king joining the Craft had encouraged many of his courtiers to also become Masons, among them were Lord Alexander, Lord Hamilton and David Ramsey,[38] who joined the Lodge of Edinburgh.[39]

When James moved down to London he continued to take part in ceremonies which involved acting out the role of King Solomon, the role taken by the Master of the lodge during the opening and closing ceremonies. And he certainly does not seem to have been secretive about it. Sir John Harrington, who spent an evening at James VI (I)'s Court while he was entertaining King Christian of Denmark in 1617 reported:

> *After dinner the ladies and gentlemen of the Court enacted the Queen of Sheba coming to King Solomon's Temple. The lady who took the part of the Queen of Sheba was, however, too drunk to keep her balance on the steps and fell over onto King Christian's lap, covering him with wine, cream, jelly, beverages, cakes, spices and other good matters which she was carrying in her hands.*[40]

This was not the only occasion on which James is reported to have carried out dramas connected with Solomon's Temple. He greatly enjoyed persuading young men to dress in flowing robes to assist him in these rituals. James formed a great affection for Robert Carr, later the Earl of Somerset, who was implicated in a murder case. Later the king turned to George Villiers, Duke of Buckingham, whom he referred to as 'his wife'.[41]

James called Buckingham by the pet name of 'Steenie'. Historian J P Kenyon comments:

> *James was head over heels in love with his 'sweet Steenie gossip', his 'sweetheart', his sweet child and wife' and a few*

> *days' absence was enough to set him throbbing with desire. 'My only sweet and dear child', he drooled, 'I pray thee haste thee home to thy dear dad by sunsetting at the furtherest and so Lord send me comfortable and happy with thee this night.*[42]

James's contemporaries were even more outspoken. Courtier Sir Anthony Weldon remarked about James's relations with a number of 'male lovelies':

> *The King's kissing them after so lascivious a mode in public, and upon the theatre, as it were, of the world, prompted many to imagine some things done in the retiring house that exceed my expressions no less than my experience.*[43]

At the time homosexuality was referred to as the vice of kings but it did not excite much public condemnation. James, however, became so obsessed with re-enacting the story of the events surrounding Solomon's Temple that his courtiers dubbed him the British Solomon.[44] But he also carried out regular Freemasonic ceremonies. William Preston reports:

> *In 1607, the foundation stone of this elegant structure [part of the Palace of Whitehall] was laid by King James, and his Wardens who were attended by many brothers, clothed in form. The ceremony was conducted with the greatest pomp and splendour.*[45]

So now I could be quite sure that James VI, through his Master of Works, William Schaw, had patronised the modern lodge system of Freemasonry in Scotland prior to his coming to England. At the time James had been initiated into Freemasonry, at the Lodge of Scoon and Perth in 1601, he had become fascinated with the rituals of Solomon's Temple, which form an important part of the Craft. James had made

Speculative Masonry fashionable in his Court in Scotland and then brought the Scottish rituals of Freemasonry to England.

It may be no coincidence that Francis Bacon, whom James greatly favoured, is shown in the frontispiece of Spratt's *A History of the Royal Society*, wearing the jewel and collar of a Chaplain of the Lodge of Edinburgh. The king had many favourites, Bacon among them, but as Bacon was not 'smooth limbed and comely' perhaps he appealed more to James's love of the rituals of Solomon. This complimented the new fashion of the society of Freemasons, which the king's Scottish courtiers brought with them to London. With Freemasonry came the study of nature and science which is the purpose of the Masonic Fellow Craft Degree.

Conclusion

Freemasonry started in Scotland, at Roslin, sometime before 1440. William Schaw, General Warden to the king of Scotland, established the lodge system around 1599. He also intended to set up Freemasonry as a Royal institution, with the King as its Grand Master Mason. With this objective in mind he cooperated with Freemason John Mylne to get the king initiated into Freemasonry at the Lodge of Scoon and Perth in 1601.

The Masons of Scotland reacted against the political aims of Schaw and Mylne and they rejected James VI as their Grand Master. Instead they affirmed their allegiance to William Sinclair of Roslin, a rather dissolute character.

James enjoyed ritual and the company of young men. He continued to practise Masonic rituals and made Freemasonry fashionable when he moved to London, as James I of England. Bacon became a favourite of James, just around the time he started to write about science.

I now knew that Freemasonry had been fashionable at the Court of James VI(I). Was this the source of Bacon's sudden interest in scientific method? John Wilkins had used Mason-speak when describing his interest in science and admitted that he had been a keen follower of Bacon.

I have shown in the Appendix that the Freemasons of the seventeenth century were studying and practising the concepts that motivated the Royal Society, and that the Masonic Fellow Craft Degree is devoted to encouraging the developing Freemason to study the ways of nature in order to 'better understand God, the creator of all'. This could have been the source of Francis Bacon's ideas.

Were the rituals and practices of Freemasonry the inspiration to avoid the discussion of religion and politics at the meetings of the Royal Society?

I decided that I should look more closely at the two best known Freemasons among the founders of the Royal Society, Sir Robert Moray and Elias Ashmole. As Moray seemed to appear at every turn of my research it seemed sensible to deal with him first. He is one of the most influential founders of the Royal Society and, despite reading all the standard histories of the Society, I still knew very little about him.

CHAPTER 6

The Life and Soul of the Royal Society

Sir Robert Moray cannot be taken to be a typical mid-seventeenth century Freemason: the fact that he reveals so much about what Masonry meant to him in itself makes him unique.[1] David Stevenson, Professor of Scottish History, University of St Andrews

I HAD REACHED A STAGE in my inquiry where I was haunted by a single shadowy figure. This man kept popping up at every turn. He seemed to have been involved in almost every key event that formed the 'Society For Promoting Philosophical Knowledge by Experiment'. He was also the driving force behind turning it into a royal club. If I was ever to understand why the Royal Society was born, I needed to know more about him. It seemed that the Royal Society was his brainchild and his influence upon it was far greater than that of any other single person.

What motivated this man? He changed sides so often during the Civil Wars it is hard to keep track of him. Moray was knighted by Charles I within days of serving as a senior member of the Army that had contributed to the king's downfall and which was still a threat to the king! He was ransomed from a Bavarian jail by the French and sent to London to act as their negotiator with the Scots. Moray helped to make it possible for Charles II to be crowned King of Scots, at Scoon, and within a few months he was imprisoned for trying to assassinate the new king. He was with Charles II in Paris when Monck decided to restore the monarchy, but he did not return to England until three months after the king. He was immediately given a grace and favour home in Whitehall and seemed to have had access to every philosopher in London within weeks. I certainly needed to know much more about this enigmatic man.

The only major biography of Robert Moray, published in 1922,[2] does not mention that he was a Freemason. However, the Earl of Elgin has preserved a long series of letters Moray wrote to Brother Mason, Alexander Bruce. [Copies may be consulted at the Royal Society.] This collection is known as the Kincardine Papers and goes into great detail about the importance of Freemasonry to Sir Robert.

Moray's life was truly extraordinary. During his sixty years he worked as a mercenary and spy for the king of France, was Quartermaster-General for the Covenantor's Army and almost managed to rescue Charles I from the Scots. Despite his apparent Royalist leanings on occasions, for example Moray led a Scots' rising against Cromwell, as we have seen he was also imprisoned for trying to assassinate Charles II. Among his eventual appointments he became Privy Counsellor, Lord Justice Clerk and Lord of Session in Edinburgh (despite having no legal experience). Moray worked as a spy for the Earl of Lauderdale and in his spare time was the life and soul of the Royal Society. Scottish Freemasonry considered him so important that they created a lodge of Research named in his honour: Lodge Sir Robert Moray, No. 1641.

Just who this man was and what drove him was a real puzzle.

Perhaps if I took all the facts I had learned about him,

arranged them in order and used the historical context of the times to illuminate his actions then I might begin to understand his motivations. However, I also needed to understand the times in which he lived, the period of the Civil War that started with the death of James VI(I).

A Martyr to the People

The backdrop to Sir Robert's story starts when the Duke of Buckingham went with the son of James VI(I), Prince Charles, on an ill-fated trip to Spain to try and win the hand of the Infanta, the Catholic king of Spain's daughter. The adventure failed, much to the satisfaction of James's Protestant subjects in England. But Buckingham then went on to negotiate a marriage between Charles and Henrietta Maria, daughter of the Catholic king, Louis XIII of France.

Three months after James had ratified the marriage treaty for his son, the first king of Great Britain died (1625) and the scene was set for the Civil War, as Charles I tried and failed to enforce his will on Parliament.

Historian Leopold von Ranke described Charles I at the time of his accession as:

> *in the bloom of life: he had just completed his twenty-fifth year. He looked well on horseback: men saw him govern with safety horses that were hard to manage: he was expert in knightly exercises: he was a good shot with the cross-bow, as well as with the gun, and even learned how to load a cannon. He was hardly less unwearily devoted to the chase than his father. He could not vie with him in intelligence and knowledge, nor with his deceased brother Henry, in vivacious energy and in popularity of disposition.*[3]

Charles had also suffered from polio as a child and spoke with a stammer.

The politics of his kingdom were quickly becoming more complex. Relying on the Royal Prerogative and the Divine right of kings was no longer acceptable to the new merchant classes. Trade was growing both within the country and overseas. Coal-mining was developing and supporting new industries. Landed gentry, grown rich on the wealth of coal-mining, were supplying Charles with the members of his Parliament. They were intent on extracting as much benefit from their legislative duties as the king could be forced into giving them and they controlled his income from taxation.

The English Protestants were anxiously watching events in Europe. They were concerned that their reformed faith would be swept away on a tide of militant Roman Catholicism. London was a thriving city, crammed with outspoken apprentices; with wealthy, self-serving city guilds and livery companies to rule them. As a result the arrival at Dover of Charles's new Catholic bride, escorted by Papist priests, was bound to set the new king off on the wrong foot with his Protestant Parliament.

Charles, encouraged by the Duke of Buckingham, his late father's favourite and his own close friend, declared war on Spain. He called a Parliament hoping they would grant him funds to fight the Catholics. His new Parliament, however, took the opportunity to review their whole attitude to taxation and decided to grant him the traditional life-long customs duties for one year only. From this first encounter Charles became dependent on regular Parliaments for his income but he greatly resented the increasing claims for more delegated authority that the newly wealthy, landed gentry of the Commons forced on him. Charles's war with Spain was intended to procure more funds from Parliament but he was offered the absolute minimum that Parliament could afford him. Then, to make matters worse, the startlingly incompetent Duke of Buckingham led an ill-fated expedition against Cadiz. It failed miserably and achieved nothing. The whole war against Spain eventually proved to be a disaster for Charles. As a result the House of Commons demanded that Buckingham be impeached and used their powers of taxation to insist the king consider their opinions on the matter:

> *We protest that until this great person be removed from intermeddling with the great affairs of State any money we shall or can give will through his misemployment be turned rather to the hurt and prejudice of this your kingdom.*[4]

Charles quickly dissolved Parliament before it could impeach Buckingham and try him for treason. He then attempted, and failed, to conclude an alliance with France against Spain. His problems were compounded by a move by the French authorities against the Protestant French Huguenots.

In 1627, in an attempt to rehabilitate his friend, Charles dispatched a naval force, under command of the hapless Buckingham, to relieve the Huguenot port of La Rochelle. True to form, Buckingham's expedition was an utter failure and Charles dared not call a new Parliament, for it would certainly have impeached Buckingham. Charles was reduced to raising money for the war by illegal means. Gentlemen were forced to lend him money and if they would not pay up, he imprisoned them. This action inflamed the landed gentry, who controlled the House of Commons, and when at last the king was forced by lack of money to call a Parliament in 1628 it was not prepared to give him a penny. Parliament demanded Charles grant them a Petition of Right before they would vote him any money at all. The king grudgingly granted the Petition but never took it at all seriously. Parliament and king were now set on a collision course.

Charles was worried about the desperate state of Buckingham's reputation. He decided to let this incompetent commander attempt to lead another expedition to La Rochelle. Its aim was to relieve the Huguenots, but before Buckingham could set sail, he was assassinated by a fanatical naval lieutenant named John Felton. Meanwhile, the Protestant Huguenots had been overrun by the Catholic French. Parliament was outraged and suspected Charles of supporting the Catholic cause. A Bill was passed declaring that anyone who furthered Popery or helped the king to collect taxes, unless authorised by Parliament, was a public enemy. Charles, in an attempt to stop the action, tried to adjourn Parliament. The Commons barred the door against Black Rod (a tradition still carried out today), and carried the Bill by acclamation. Charles, in desperation, dissolved Parliament in 1629.

He made the following announcement:

> *We have showed by our frequent meeting our people, our love to the use of Parliaments; yet, the late abuse having for the present driven us unwillingly out of that course, we shall account it presumptive for any to prescribe any time unto us for Parliaments, the calling, continuing, and dissolving of which is always in our own power, and shall be more inclinable to meet in Parliament again, when our people shall see more clearly into our interests and actions and when such as have bred this interruption shall have received their condign punishment.*[5]

For the next eleven years Charles ruled without calling a Parliament. He continued to collect customs, without the authority of Parliament, and also imposed a new nationwide Ship Tax to fund the costs of running the navy. It is certain that he had strong political and military justification for this tax. The fleet was in a bad state of repair and the Barbary corsairs were a constant threat to the western coasts. In addition the Dutch, French and Spanish warships were continually flouting England's sovereignty of the Channel and the North Sea, and sailing within English territorial waters with impunity. Many people objected to this new tax but Charles simply imprisoned them. He also introduced property taxes and a tax on failure to attend the established Church every Sunday.

Unfortunately, Charles I had been crowned King of Scots at Holyrood House in Edinburgh, in the presence of Archbishop Laud of England. This officiation at a Scottish coronation by an English bishop was seen by the Scots as an attempt by the Episcopal Church to move into their kingdom. It caused some rumblings of discontent amid the Presbyterians of Scotland, and they were outraged when Charles, possibly with a view to extending his lucrative church attendance tax, ordered that the English Prayer book was to be used in Scottish Churches henceforth. This new Prayer Book, foisted on Scotland in 1637, was too closely associated with the hated Archbishop Laud and the Episcopalian Church to be acceptable and the Scots would not support it. A date was set for the first public use of the new Prayer Book on Sunday 23 July 1637 at St Giles Church in Edinburgh. The people of Edinburgh packed St Giles that day but not to listen to the readings.

Legend credits a cabbage stall holder called Jenny Geddes with starting the riot when she threw her stool at the priest who was about to read from the English Prayer book. Whether she did or not, the whole of the vast congregation forcibly ejected all the Episcopal priests and their new Prayer Book from the Church before conducting a Presbyterian Service.

Charles had lived in England too long and seemed to have forgotten the independent nature of his Scottish subjects. Attempting to force the Episcopalian form of worship upon them was a good example of his ignorance of Scottish sensibilities. The Scots did not intend to accept this interference in their freedom to follow their own beliefs and decided to do something about it. On 28 February 1638 the Presbyterians, lead by the Earl of Sutherland, held a meeting in Greyfriars' Kirkyard in Edinburgh where they signed a document which reminded their king, in far away London, that his subjects in the Kingdom of Scotland were not English and were not to be treated as a part of England. The document, which became known as the Covenant, was signed by all those assembled at Greyfriars' and later by many more supporters. The Covenant promised 'to defend the worship of the forefathers against the king'. All those who assented to this action, among whom were many noble families of Scotland, were known as Covenantors.

The king was now faced with a highly organised and very hostile General Assembly of the Covenantors who were demanding the abolition of the Episcopacy, headed by Charles. Charles, however, was always insistent on getting his own way. He simply bided his time. He intended to muster an army to force his Scottish subjects to obey his command. Charles ordered the Assembly to disband and it refused. When the Covenantors realised what Charles was intending to do they also formed an army and marched South to meet the forces of the king. The matter came to an armed confrontation at Dunse Law, in Berwickshire, in May 1639. It is known to history as the First Bishops' War and the king lost it. He backed down and agreed to call both a General Assembly of the Church of Scotland and a Scottish Parliament. This show of public opposition by the Scottish people had forced the king to rethink and Charles yielded; the new Prayer Book, the bishops and the wearing of surplices were all done away with.

Charles, however, still wanted his will carried out and when the Covenantors returned to Edinburgh they found that he had issued a new order demanding the practices of the Episcopalian Church to be adopted in Scotland. This time he intended the command to be enforced by the Earl of Stafford. Stafford had already subdued Ireland (in the process creating an Irish standing Army) and was considered to be a hard enough man to take control of the rebellious Scots. All he needed was an English army, but Charles couldn't afford one without the financial support of the English Parliament. Stafford suggested bringing his Irish troops to England but Charles was too afraid of an English backlash. So on 13 April 1640, his shortage of money forced him to call a new Parliament in London.

This new Parliament refused to grant Charles any money and so he was again pressed into illegal taxation, enforced by the brutal Stafford. Now the king found the forces of Parliament were ranged against him. Under the mild despotism of his period of personal rule the landed gentry had established strong control of local government in many parts of the country. They formed the Parliament and they were unhappy with a king who denied their property rights and taxed them unlawfully. Eventually they came to rely on a strong leader from among their midst. His name was John Pym.

The Covenantors marched down as far as the River Tyne where they faced the ramshackle assembly of conscripts, which was the only army that Stafford had been able to raise. Parliament and the Puritans of England supported the Scots and the English army had little stomach for a fight. After a few days stand-off the English fled and the Scots took Newcastle. Sir Winston Churchill describes the battle thus:

> *The Scots cannon fired and all the English fled. Never have so many run from so few with less ado. At this moment King Charles's moral position was at its worst. He had plumbed the depths of personal failure.*[6]

Charles had no choice but to deal with John Pym's House of Commons. Pym drove a hard bargain and the Earl of Stafford's head was duly delivered at Tower Hill on 12 May 1640.

Charles then went to Edinburgh to try to make his peace with the Scots. He assented to the establishment of total Presbyterianism in Scotland but was accused of complicity in the kidnap of the Marquis of Argyll. As a result the Scots rejected him. They insisted the king should pay for the support of their Army in Newcastle. (Stafford's Irish army had been disbanded on his death.) Now Ireland also rose against the king and by harshly suppressing the Irish the king lost the trust of the English as well.

Parliament produced a document called the Grand Remonstrance that set out their grievances against the king. This document so annoyed the king that he decided to arrest the five ringleaders among his opponents. Charles, accompanied by three or four hundred troops, went to the House of Commons on the afternoon of 4 January, 1642. He entered the House and demanded the five members be handed over, but they had already escaped. When he left the Commons he was mobbed by Londoners, outraged that he had breached the privilege of the House of Commons. Charles was forced to flee to Hampton Court.

Hostilities ensued, which lasted until 1646, when Royalist Oxford fell and the king eventually surrendered to the Scots army at Newark so ending the First Civil War. For a year Charles was imprisoned by the Scots at Newcastle and then handed over to Oliver Cromwell's Parliament, in return for back pay to disband the Scots Army. For a while it looked as though Charles would come to a working arrangement with Cromwell, but instead he signed a deal with the Scots to introduce Presbyterianism as the established religion of England. This action provoked the Second Civil War.

The Second Civil War was simple and short. The king, the Lords and Commons, the landlords and merchants, the City of London and the countryside, the Scots Army, the people of Wales and the English Fleet faced Cromwell's New Model Army. Cromwell soundly beat them all and by the end of 1648 was Dictator of Britain. Charles was left with only Carisbrooke Castle. Cromwell had the king taken back to London. There he was tried and found guilty of treason. The king was beheaded on 30 January 1649.

This then was the background against which Sir Robert Moray had grown to manhood and the events which had shaped his career. Knowing this I was ready to look more closely at the man himself.

Unveiling the Unique Freemason

Robert Moray was born the son of Perthshire Laird Sir Mungo Moray of Craigie, on 10 March either in 1608 or 1609 (nobody is quite sure). Most of what I ultimately learned about his early life came from comments he made himself, in a long correspondence he held with Alexander Bruce, while they were both exiled during Oliver Cromwell's rule. Moray's mother was the daughter of George Halkett of Pitfirran. The Moray and the Halkett families were resolute supporters of the Stuarts both before and during the Civil War. One of Robert's cousins, Anna Moray, helped the young Duke of York escape to Holland in 1647, by disguising him as a young girl.[7] At the time she was the lover of Joseph Bampfield, who was another of the negotiators between Charles I and the Scots Covenantors. Once young James was safely abroad Anna had returned to Edinburgh, and assisted in the care of the wounded after the Battle of Dunbar. She is reputed to have been skilled in surgery. Later, she became Lady Anna Halkett when she married Sir James Halkett, a grandson of George Halkett. From then on she lived a very quiet life, surviving until the end of the century, passing her time writing devotional books and her autobiography, which was published posthumously in 1879.[8]

Robert's curiosity about technical matters was first aroused at the age of fifteen. As he explained in a letter to Bruce, he was taken on a visit to 'the moat at Culross, when the coal was going there'.[9] This moat was an artificial island that had been created in the mud flats of the tidal reaches of the Firth of Forth. From this island a shaft had been sunk, to enable coal to be mined. To create the working mine had involved solving many problems of construction, waterproofing and pumping. The design work involved fascinated Robert and this is probably the inspiration of his life-long interest in civil engineering.

Sir George Bruce's underwater coal mine was one of the wonders of Scotland in 1623 when young Moray went to see it. It was situated on the Fife shore of the estuary, on the seaward side of the present Kincardine Bridge, where the A876 crosses the Firth. The first tunnels had been cut in 1590 and by the time young Robert walked through its arched walkways it extended over a mile out below the Firth. The mine was an ingenious structure that had been built by first creating a strong circular wall of stone on the beach at low tide. This wall had been waterproofed with mud and bitumen to make it secure. Sir George's miners then dug down, first through the artificial island and on through the silt of the seabed and its underlying rock. They kept digging, with their mattocks and picks until they reached a layer of sea coal, forty-five feet (fifteen metres) below the surface of the sea.

Seepage of sea water into the shaft was a problem. Robert was impressed to discover that this had been solved using a complex mechanical contraption. A downward sloping passage had been cut back to the sea shore and a system of drains dug which all flowed back to a sump, which was lower than the level of the mine. This alone would not have kept the mine dry as it would in time have filled up. Above the sump was a donkey engine, driven day and night by three horses. Each horse would be harnessed to the pole and then plod around for eight hours, before being given sixteen hours rest. The donkey engine turned a large necklace of iron links with thirty-six buckets attached to it. It was carefully balanced so that as eighteen buckets went down the sump, the other eighteen buckets came up. The horses only lifted the weight of the bucket's contents, the lifting mechanism being finely balanced. Even from this mechanism Sir George made a profit. The pumped sea water was fed back to the sea via a series of evaporation pans that yielded up to one hundred tons of salt a week. Sir George not only supplied much of Scotland with salt and coal, he also exported salt to England and Germany, from the adjacent port of Grangemouth. Was this where Robert Moray first learned how combining necessity with inspiration could solve problems profitably?

Hume Brown, writing in 1618, said of this wonder mine:

> *The mine hath two ways into it, the one by sea and the other by land; but a man may goe into it by land, and returne the same way if he please, and so he may enter it by sea, and by sea he may come forth of it: but I for varieties sake went in by sea, and out by land. Now men may object, how can a man goe into a mine, the entrance of it being into the sea, but that the sea will follow him and so drown the mine? To which objection thus I answer, that at low water, the sea being ebbed away, and a great part of the sand bare; upon this same sand (being mixed with rockes and cragges) did the master of the great worke build a round circular frame of stone, very thicke, strong and joined together with glutenious and bitumous material, so high withall, that the sea at the highest flood, or the greatest rage of storme or tempest, can neither dissolve the stones so well compacted in the building or yet overflowe the height of it.*[10]

The ingenuity and wonder of this magnificent engineering construction stayed with Moray all his life and encouraged him to study ways of building strong structures.

When he was twenty years old, Moray became interested in philosophy, particularly a type known as Christian Stoicism. It was a reworking of the old Stoic idea that 'ethics are the most important area of knowledge'. Robert's concept, however, encouraged him to look to logic and natural science for ways of explaining ethical beliefs. He soon became obsessed with developing ways of keeping his feelings and emotions under rigid control. Even during the horror of his wife's fatal childbirth, observers marvelled at his unyielding stoicism, a control which he did not allow to crack even while comforting Sophia during her death agonies.[11] It is this reputation for detached coolness that made it so hard to discover anything about the true personality of this aloof and distant man. The only passion Moray ever seems to display, after the death of Sophia, is when he is writing about science or Freemasonry.

Professor Stevenson, whose quotation opens this chapter, has made a detailed study of Moray. He believes that by his late twenties it would be correct to describe Moray as an engineer.[12] Moray confirms this from his correspondence, saying that in 1637 he was seeking out the company of engineers in Islington who 'pretended great skill in aqueducts'.[13] By then, Moray was a full-time soldier. In 1633, at the age of twenty-five, he had joined an elite Scottish Regiment that served as bodyguard to the King of France. They were based in Paris and known as the Scots Guard.[14] While serving with them he became a favourite of Cardinal Richelieu, who sent him on regular missions to monitor the relationship between Charles I and the Scots. His visit to London in 1637 was one of these intelligence-gathering trips.[15] Historian Patrick Gordon, said of this arrangement:

> *After he [Cardinal Richelieu] sounded the depth of the man's mind and finding he [Sir Robert Moray] was indifferent, so as he could make a fortune, whether it were with the King or with the malcontented Puritans, he finds no difficulty to persuade him that his love for the Scots, by virtue of their ancient league made him lament their cases.*[16]

One of the Cardinal's pet projects was the *Académie Française.*

Richelieu had fostered a series of regular meetings between men of letters in Paris. He later formed these men into an authoritative body to address any questions concerning the language, and literature, of France. Richelieu established this Academy in 1635.[17] The *Académie's* success, after receiving royal approval for its work, could well have inspired Moray. Twenty-five years later, he also would seek a Royal Charter of approval for the new scientific society he had conceived.

In 1639, when the Covenantors were starting to flex their military muscle in Scotland, they had warned Charles I that if he planned to impose the Prayer Book on Scotland he would need 40,000 men to do it. Edward Hyde, writing at the time said:

> *a small, scarce discernible cloud arose in the North, which was shortly after attended with such a storm, that even rooted up the greatest and tallest cedars of the three nations; blasted all its beauty and fruitfulness; brought its strength to decay and its glory to reproach.*[18]

Richelieu quickly spotted this little cloud. He saw an opportunity to promote the interests of France, by supporting the Scots against Charles I. Moray was promoted to Lieutenant-Colonel in the Scots Guard and dispatched to Scotland. His instructions? To assist the revolt against Charles. Patrick Gordon traced Richelieu's plans:

> *Wherefore, Cardinal Richelieu, choosing forth a man fit for his purpose amongst the many of the Scots gentry that haunted the French court, he chooses forth one, Robert Moray, a man endowed with sundry rare qualities, and a very able man for the Cardinal's project.*[19]

The Covenantors had been fortunate in obtaining the service of a very experienced soldier to lead their army. Sir Alexander Leslie had just returned to Scotland after spending thirty years in the Army of the King of Sweden. The 60-year-old general signed the Covenant and was immediately placed at the head of the Covenantors' Army.[20] Leslie saw Moray's engineering skills as a distinct military asset and promptly gave him the post of Quartermaster-General. The duties of the Quartermaster-General were to assign quarters and to supply weaponry to the army. He was also in charge of building, and laying out fortifications and camps. Moray's knowledge of surveying, mathematics and civil engineering was invaluable to Leslie.

Alexander Hamilton was the Covenantors' general of artillery. He worked closely with Moray and introduced him to Freemasonry.

Hamilton was a member of the Lodge of Edinburgh, having been initiated in Edinburgh on 20 May 1640.[21] Alexander Hamilton in turn had been initiated by James Hamilton, and John Mylne. The networking between the Scottish nobility was starting to become interesting! James Hamilton was the third Marquis of Hamilton and at the time was supporting Argyll. He had been Charles I's advisor on religion but had resigned in 1638. James Hamilton served with the Covenantors' Army against Stafford, although he would not sign the Covenant himself. Eventually he left Leslie and in 1643 went to Oxford, to try to join the king and persuade him to do a deal with the Scots.[22] John Mylne, the grandson of the Mason who initiated James VI into the Lodge of Scoon and Perth, had himself initiated Lord Alexander of Stirling, Charles I's Secretary of State for Scotland, into the Lodge of Edinburgh on 3 July 1634. The following July he had initiated Anthony Alexander, Charles I's Master of Works. On 27 December 1637 he had initiated David Ramsey, Royal clockmaker, and gentlemen of the Bedchamber for Charles. John Mylne was certainly well connected with the Scottish nobility and when he initiated

Alexander Hamilton he began to extend his links into the officers of the Covenantor Army.[23]

By May 1641, Moray was encamped in Newcastle, with the rest of Leslie's Army. He was approached by General Hamilton, and asked if he would like to become a Freemason. Hamilton must have explained that as well as himself the Marquis of Hamilton was already a Mason and would be conducting the ceremony alongside the Army's master builder, John Mylne, who would have reported to Moray in his position as Quartermaster-General. By the end of the evening Moray would have known that many Scottish noblemen were Freemasons. Moray agreed to join and on the evening of the 20 May 1641, James Hamilton, Alexander Hamilton and John Mylne, brought together at least four other Freemasons and convened a lodge to initiate Robert Moray. A minute was produced of the initiation and returned to the Lodge of Edinburgh.

Comparing the signatures of James Hamilton and John Mylne they are clearly the same as those on the record of Alexander's initiation, exactly a year earlier. Alexander Hamilton had signed with his Mason's Mark, an equilateral triangle. Likewise Robert Moray has signed his name with his Mason's Mark, a five-pointed star. He was now a Fellow of the Craft, and so entitled to his own Mason's Mark.

Imagine the voice of the Marquis of Hamilton telling him, 'Brother Moray, we congratulate you on your preferment . . . you are to judge with candour, admonish with friendship and reprehend with mercy.' I was to discover that he remembered and applied these sentiments for the rest of his life. But the phrase that had most impact on his future behaviour was, 'you are to contemplate the intellectual faculties and to trace them in their development through the paths of nature and science even to the throne of God Himself'.

This idea would grow in his mind until it finally blossomed in the form of the Royal Society. In one of his letters to

Alexander Bruce, written just before the Restoration, he said:

> *Many such things have befallen me in my life which have given me so intimate an acquaintance with God's goodness in such dealings, that I have much cause to thank Him (besides the good I am indebted to Him in the several dispensations), for stooping so far as to give me so many pregnant sensible experiments for confirming my faith and His Truth.*[24]

Robert Moray left the Covenantors' Army, at the end of 1642, and travelled south. On 10 January 1643, he attended the Court of Charles I in Oxford. I was aware of this fact because it was on that day Charles knighted him. Why Charles chose to knight Moray had puzzled me for quite a while as it seems a very strange turn of events. Moray's mission had been to promote the Scottish cause in order to make trouble for England to the advantage of France. He would not at first sight seem to be an obvious candidate for the favour of the beleaguered king of England. But there was another factor to take into account. Cardinal Richelieu, Moray's patron and spymaster had died on 4 December 1642. Charles knighted Moray exactly thirty-seven days later. There was adequate time for Moray to have received the news of Richelieu's death. Now that he had lost his French patron, was it possible that Moray had decided to change sides? Indeed, was Robert Moray the messenger who told Charles of the death of Richelieu? And is this why he travelled down from Newcastle to Oxford?

Richelieu's death would have allowed Queen Henrietta Maria's brother, the 41-year-old Louis XIII, to abandon Richelieu's Anti-Huguenot policy and make it possible for him to support his sister's husband in his battles with a Puritan Parliament.[25] Moray still held a commission as a Lieutenant-colonel in Louis's personal bodyguard when he returned to Paris soon after receiving his knighthood. Did he carry a message to Louis from Charles? I mentioned this scenario in a conversation with Robert Cooper, the Librarian of Grand Lodge in Edinburgh. Robert added the comment that if Charles used Moray as a messenger then he certainly would have knighted him, to make certain his message was given proper status by his French brother-in-law. I cannot prove this hypothesis but the fact that on his return to Paris Sir Robert was immediately promoted to a full colonel in Louis's personal body guard tends to support the idea that his status had been deliberately increased in order that he could carry an important message to the French king. It was unfortunate for Charles that Louis died soon afterwards and so was not able to assist him.

Towards the end of 1643 Moray returned to active duty with his regiment of Scots Guard, fighting in Bavaria.[26] He was unlucky enough to be captured and imprisoned, and spent the next fifteen months sitting in a Bavarian prison cell, studying magnetism and corresponding with the German scientist Kircherus. Then, after nearly two years of imprisonment he was suddenly ransomed by the French. On 28 April 1645, Sir Robert was freed, on the payment of £16,500 by Cardinal Mazarin.

Sir Robert returned to a France which again wished to exploit the differences between king and parliament in England. Louis XIII had died soon after Sir Robert had been imprisoned. France was now ruled by Queen Anne of Austria (mother of the young Louis XIV) and Cardinal Mazarin.[27] Charles I had lost the city of Bristol and, having few options left to him had opened up discussions with the Covenantors. Moray's old Commander, General Alexander Leslie, now the Earl of Leven, was the main negotiator for the Covenantors.

Sir Robert returned to Paris after Cardinal Mazarin unexpectedly paid his ransom.[28] Perhaps Mazarin remembered that Charles had favoured Sir Robert enough to knight him. Could that trust be turned to French advantage? Mazarin immediately dispatched Moray to London. There he acted as an intermediary between the French Ambassador Montereul, King Charles and General Leslie. He took part in the negotiations with Charles, which were difficult and protracted. Did Mazarin decide to ransom Sir Robert so he could take up his old role as a French *agent provocateur* between the Scots and the English? If Mazarin was Moray's new spy-master I wondered what instructions he might have given to Sir Robert.

Sir Robert accompanied Charles when he decided to surrender himself to Leslie at Newark and travelled on with the king to Newcastle.[29] He did nothing further until the Covenantors decided to sell Charles to Parliament for the £400,000 back pay owing to their Army.[30] It was at this point Moray tried to arrange for Charles to escape to France. Had he been ordered to change his position by Mazarin? On the evening of 26 June 1646 Charles, Prince of Wales, had landed at St Malo en route to join Queen Henrietta at St Germain. Henrietta, who was now living as an adjunct to the French Court, had plans to marry her eldest son to Anne-Marie-Louise de Moutpensier, a cousin of the young Louis XIV and a very rich lady, known as La Grand Mademoiselle. This move to France, creating the possibility of allowing himself to be used as a pawn by Mazarin, was a major political mistake on the part of the Prince of Wales[31] but it had the immediate effect of encouraging the French to try and get the English king to France as well. This may have been why Moray, as their Scottish agent, suddenly decided to help the king to attempt an escape.

On Christmas Eve 1646 when Moray's plan to bring Charles I into French control had failed Charles was taken into custody by Parliament. Sir Robert went to Edinburgh, possibly to recruit more Scottish gentlemen for the elite Scots Guard of the King of France. On 27 July 1647 the minutes of the Lodge of Edinburgh show him to have been present at the initiation of

William Maxwell and another unnamed gentleman. In the first part of the seventeenth century, Edinburgh Lodge had been fashionable with the Scottish nobility. It had also appealed to the courtiers of Charles I, including the Dukes of Hamilton and the Earls of Stirling. As a result it would have been a useful place for Moray to renew old contacts and network with the Scottish leaders.

By May 1648 Moray was back in Paris. He had been asked by John Maitland, the Earl of Lauderdale, to try to persuade Charles, Prince of Wales, who was then in Paris,[32] to come to Scotland and lead an uprising by a group known as the Engagers.[33] Lauderdale's choice of Moray for this job might well have been a direct consequence of Moray renewing his contacts with his Masonic Brethren of the Lodge of Edinburgh. This is likely because one of Lauderdale's fellow Royalist conspirators was James Hamilton, the Freemason who had initiated Moray.

However, it does not really matter why Lauderdale chose Moray because this first contact with the young man who would become Charles II was to change the direction of Sir Robert's loyalty for the rest of his life. When Charles decided to go to Scotland, Moray, then forty-two, resigned his commission in the Scots Guard and also returned to Scotland. I was interested in investigating Moray's dealings with Charles II, both as Prince of Wales and King of Scots. Was this relationship where the real foundations of the Royal Society were laid?

After the beheading of Charles I the Scots offered the crown of Scotland to his son, provided Charles II signed the Covenant. At first Charles II was unwilling to consider these terms because he thought he might still retake Scotland by force. But what part had Sir Robert played in these events?

The Supporter of King Charles II

While Charles was still Prince of Wales and living with his mother he developed a way of dealing with hysterical women which he would later extend to a method of coping with the conflicting demands of the various supporters he had to satisfy. Historian Hester Chapman describes this technique as:

> *agreeing to everything they [his womenfolk] said, granting all requests and then going his own way. His courtesy to Henrietta Maria never failed; he remained respectful, protective, and affectionate; but he made his own decisions and stuck to them.*[34]

This is a method he would develop to a fine art in Scotland when he was forced to deal with the Covenantors. Early in May 1648, before the death of Charles I, Moray delivered to the Prince of Wales a formal letter from Hamilton and Lauderdale formally requesting the Prince to come to Scotland to lead a group of Stuart supporters. On 30 May Prince Charles wrote back saying he was:

> *inexpressibly desirous of himself and impatient to be amongst them.*[35]

Acting as a go-between for Lauderdale and the Prince of Wales, Moray set up a meeting to be held at Helvoetsluys, in the Netherlands. Soon after agreeing to this meeting Charles heard that the bulk of the English Fleet had deserted Parliament and were sailing to his support. On 25 June 1648, Charles left Paris for Calais. He now saw an opportunity of using the English Navy and Scots Army to restore his father's kingdom by force. Joining the fleet at Calais, Charles sailed to his meeting with Hamilton aboard HMS *Satisfaction*, arriving in Helvoetsluys on 9 July 1648.

There were now three distinct political factions in Scotland; the Royalists, led by the Marquis of Montrose; the Hard Line Covenant party led by the Marquis of Argyll; and the

Lauderdale-Hamilton Faction known as the Engagers, which supported Presbyterianism but did not like the overriding power of the Kirk which Argyll supported. Although Hamilton and Lauderdale had signed the Covenant they were now prepared to work against it as they had decided the balance of power between the Clergy and Nobles had swung too far towards the Preachers.

Lauderdale and Hamilton had a majority in the Scottish Parliament, but did not have the overall support of the people of Scotland. They had already been involved in attempting to 'do a deal' with King Charles I twelve months earlier, while he was imprisoned on the Isle of Wight.[36]

Neither side had the Prince of Wales's interests at heart. Argyll wanted to defeat the Engagers and used the deal with King Charles I as evidence they were prepared to sell out the Kirk. Lauderdale and Hamilton wanted to use the Prince of Wales to weaken Argyll's support among the common people of Scotland. At this stage in his career, Sir Robert was not acting in the best interests of the Prince of Wales, but he was now acting in what he saw as the interests of the people of Scotland. Moray believed in religious tolerance, which the Kirk did not.

Negotiations with Lauderdale were continuing when the Prince of Wales set sail for Scotland on 17 July. Hamilton had already started to march south with a large force of Scots, although Argyll's hard-liners had not joined him. He crossed the border into England on 8 July 1648. The Scots now desperately needed the Prince of Wales at the head of their Army, so that it could be portrayed as a Monarchical Army of Liberation, rather than a 'plundering, ravaging rabble of Scots intent on pillaging anything they could in England'.[37]

Lauderdale has the distinction of being one of the only two Scotsmen that Charles II ever liked (the other being Sir Robert). Prince Charles wrote of Lauderdale:

We are like to be very happy with him.[38]

But despite this goodwill Prince Charles had not immediately headed towards Scotland with his newly acquired fleet of eleven ships. He blockaded the Thames for a while and even considered attempting to take the Isle of Wight and rescue the king. He was not yet prepared to make a deal with Lauderdale because it involved embracing Presbyterianism. Charles argued the finer points of religious agreement until 16 August, before he finally accepted Lauderdale's terms. The following day Cromwell wiped out Hamilton and the Engagers' Army at the battle of Preston. The opportunity for retaking England was gone and Charles now had to deal with Argyll and the Hard Line Covenant Fundamentalists. To make matters worse, Prince Charles's over-enthusiastic sailors threatened to throw Lauderdale overboard. Charles's biographer Antonia Fraser said of this sad sequence of events:

The whole incident left a bitter taste behind it. Each party, Royalists and Scots, could argue that they had been let down by the other. This sour sentiment was not a good omen for any future co-operation between them.[39]

Lauderdale reported back to the Scottish Parliament:

It was impossible to obtain more in religion from the Prince.[40]

Charles returned to exile in Holland and Sir Robert went back to Edinburgh with Lauderdale. In the meantime Cromwell executed Charles I.

Saved by the Mist

Legend has it that on the afternoon of 5 February 1649 Charles II learned of the death of his father when his private chaplain,

Stephen Goffe, addressed him as 'Your Majesty'. But he was a king in name only, except in the Island of Jersey, where he was proclaimed eleven days later. King Charles II had few options left to him. If he wanted to try and claim his kingdom, Scotland was his only practical possibility. The Scots were by and large still monarchists, but possibly only because they were not prepared to allow the English to abolish their monarchy. When the news of the execution of Charles I reached Edinburgh Charles II was proclaimed king of Scotland at the Mercat Cross, but this was not the same as actually being crowned. Could Ireland be persuaded to provide the troops to overcome Cromwell? Queen Henrietta, helpful as ever, wrote to Charles telling him immediately to embrace the Catholic religion so that Ireland would rise and restore him to the throne.[41] Cromwell seemed to agree there might be something in this idea as he promptly abolished the monarchy and sold all the remaining property belonging to the Crown. Charles opened a new set of discussions with Argyll's fundamentalist Covenantors, but Argyll said he would only assist if the king took the Covenant. Charles was not yet desperate enough. He still had hopes of Ormande in Ireland and Montrose in Scotland.

For a while Charles was encouraged by the success of the Duke of Ormande in Ireland. But by 15 March Cromwell was preparing to invade Ireland and destroy the king's supporters there once and for all. The Marquis of Montrose had managed to raise an army and landed on Orkney. From there he started to march southwards through Caithness. Argyll sent an Army of Covenantors against him and defeated him at Invercarron. When he heard this news Charles agreed with Argyll that he would sign the Covenant, in return for being crowned King of Scots.

Charles now adapted his technique for handling women to handling politicians. He was prepared to agree to anything but kept his own council. He had little choice in the matter. If he was not prepared to accommodate even those parts of his kingdom which wanted to crown him king, then why should anybody else support him? Although his experiences in Scotland were disastrous at the time, had he not tried to win back his kingdoms, Monck may never have considered Charles a possibility for restoration after the death of Cromwell. It was not an easy course for Charles. Argyll and his black-robed, hard-faced ministers of the Kirk did not intend to spare him any conceivable humiliation.

Charles sailed for Scotland on 24 May 1649. The voyage took many long days, tacking against the opposing winds, miserable with sea sickness, hunted by Parliamentary warships and plagued by increasingly stringent demands from Argyll. On 3 July, while still on board ship, he signed not only the National Covenant but also the later Solemn League and Covenant, which was intended to make Presbyterianism the religion of all Britain. The Covenantors must have believed that God was on their side, for no sooner had Charles signed than the wind dropped and the weather appeared to help him at last. As his ship tried to slip past Cromwell's warships into the mouth of the River Spey, the *Haar* came down. The *Haar* is a deep, damp, impenetrable mist which sometimes forms during the summer months on the East coast of Scotland. Under cover of the cold iron hand of the God of the Kirk, Charles made safe landing, saved by Scotch mist.

The Old Rituals of Kingship

Where was the 42-year-old Sir Robert while all this was taking place? For once he was giving way to his emotions and revelling in courting 25-year-old Sophia, the daughter of Alexander Lindsey, Earl of Balcarres.[42] Balcarres, who had signed the first covenant, fought at the Battle of Marston Moor on the side of Charles I, but had joined the Lauderdale-Hamilton Engagers in 1648 when he decided that the Kirk was becoming too extreme in its views.[43] As Sir Robert was now working closely as a negotiator between Charles II and the Engagers, I can only guess that through them he came to meet Sophia.

I already knew, from his letters, that Sir Robert had developed an interest in symbolism and Egyptian hieroglyphics. When he met Alexander Lindsey, who was ten years his junior, he found a man who was to introduce him to the occult lore of Rosicrucians. Balcarres had an extensive collection of books and manuscripts concerning alchemy and the brotherhood of the Rosy-Cross.[44] Gould's *History of Freemasonry* refers to 'Sir Robert as a great patron of the Rose-Crusians, an enthusiasm he shared with Brother Mason Elias Ashmole'.[45] This is probably because Moray eventually inherited Balcarres's library which he in turn left to the Royal Society.

Meanwhile Charles was courting the Marquis of Argyll and not enjoying it very much. The God John Knox had envisaged, had become an all-devouring killjoy. Laughing, dancing, singing, playing games or any other pastime enjoyed by the young was considered a sin. The seduction of young women was perhaps the one sin Charles missed most of all. When he first stayed in Scotland he was lodged in Dunfermline and he managed to occasionally receive some of his Engager supporters, occasionally with their ladies. One day, well before his Coronation, he had just finished listening to a long lecture on his shortcomings delivered by Robert Douglas, the Moderator of the Kirk and a number of his fellow invigilators of sin. Charles had received them in his bedchamber, as a mark of respect, and he was still in his bedchamber as the black-clothed clerics left him. He remained there when one of his loyal supporters and his new young lady were shown in to meet him. As the Dictators of the Kirk were passing below his window he was observed to be congratulating the young woman on some good fortune and kissing her most enthusiastically. The ministers wanted the king to be publicly reproved, lectured and then made to serve a period on the penance stool. Douglas objected, arguing that to expose and degrade the man they were soon to crown king, would reflect badly on themselves. He won the day and was allowed to call the king to account privately. He did so by quoting an old Scottish proverb. 'When one is inclined to kiss his neighbour's wife, it is proper to shut all the doors and windows.'[46] This incident occurred around the time Sir Robert was courting and about to propose to Sophia Lindsey. Balcarres, her home, is quite close to Dunfermline and a week or so earlier Charles had granted an audience to Sir Robert's cousin Lady Anna Moray and had spoken kindly to her.[47] It is nothing more than a plausible romantic notion but I like to imagine that the lady King Charles was congratulating with a kiss when he was caught out by the Kirk, might have been Sophia, on the occasion of her engagement to Sir Robert.

By now Cromwell had returned from Ireland and was advancing at the head of his New Model Army towards Scotland. General David Leslie was in command of the Scottish troops and morale was good, with the king now firmly trapped within the politics of the Kirk and Covenantors. So it came as a tremendous shock when Cromwell, on 3 September, defeated the Scots at the Battle of Dunbar and left 3,000 of their finest soldiers dead. The Kirk responded by appointing a special feast day to bewail the sins of the entire Stuart family, but it was too late! The army Charles had hoped to use to regain England had become a spent force, but Charles hadn't yet realised it. If he was to have any chance of retaking England he would have to bring all the squabbling Scottish factions together. This was proving almost impossible.

When he signed the Covenant, Charles had expected his coronation to take place at Holyrood, in Edinburgh on 15 August 1650. Cromwell's invasion prevented that. Charles needed to be crowned but he was growing weary of the Ministers of the Kirk and their continual rehearsal of his failings. He later said of this time:

> *It was a miserable life. I saw no women and the people were so ignorant that they thought it sinful to play the violin.*[48]

Argyll then decided to make the best of a bad job and to take advantage of not being able to crown Charles at Holyrood. The traditional crowning place of the King of the Scots, since the times of King Malcolm Macalpine, had been at Scoon, just outside Perth. Now a virtue was made of necessity and great play was made by attempting to focus nationalist sentiment on Charles's lineage and his right to be crowned at Scoon. His father had been the first King of Scots since 1296, who had been crowned while seated on the stone of Scoon. The traditional coronation stone which had once been used in the ceremonies at Moot Hill had been removed to Westminster Abbey in 1296 by Edward I of England.

It was not possible for Charles to be crowned in Westminster, seated on the Stone of Scoon, but it would be possible to hold his coronation at Moot Hill, near Perth, making use of the ancient rituals of Scottish king-making.

The missing coronation stone could only add to the poignant symbolism of the ceremony. Once Charles was King of Scots he would retake England and the ancient stone Cromwell now held. Historian R J Stewart commented on this king-making stone and the power it conferred:

> *The stone upon which British rulers are crowned, is the one carried off from Scotland by Edward I in 1296. This is [possibly] the original Stone of Scone, used for the installation and crowning of Scottish kings; some writers maintain that this stone was the sacred king stone from Tara in Ireland, brought to Scotland by the Dalraidic kings. Magically it is essential for the royal line of Britain to be installed upon the*

sacred stone, a tradition still upheld today.[49]

To this day the kings and queens of England have to be crowned while sitting above the stone of Scoon. Historian of heraldry W J Bennett commented on the power of this stone when creating a legitimate king:

The jewel studded regalia takes second place to that piece of rough and apparently valueless stone on which the sovereign sits for the actual crowning. No doubt a desire to perpetuate an ancient custom could account for the continued use of this 'Coronation Stone', but this does not explain the origin of this custom; why this particular stone was chosen or the veneration in which it has always been held by people of the British race.

According to tradition . . . for nearly a thousand years, the kings of Ireland were crowned while seated on it. It was then taken to Scotland and used for the same purpose until Edward I took it to Westminster.[50]

What is certain is that the Scots had a tradition of holy stones which was part of their king-making rituals, and this tradition included Moot Hill, in the grounds of the Palace of Scoon. There was no chance of Cromwell offering the Stone of Scoon to Argyll for Charles II's Scottish Coronation; Oliver was too busy invading Scotland to be bothered by such niceties. His attacks on Edinburgh made it impossible for Charles to be crowned at Holyrood, but the traditional crowning place of the King of Scots was available. The symbolism of crowning Charles on the same mound where King Robert the Bruce, the victor over the English at the Battle of Bannockburn, had been crowned would not have been missed by Argyll. Using Scoon as the site of Charles II's coronation, as King of Scots, would help focus the support of the Scots people on him and strengthen

Argyll's position. Accordingly Charles and his court were set up at 'the King's Great Lodgings in Speygate', at Gowrie House in Perth.

Following in Charles's Footsteps

I decided I would follow in the footsteps of Charles II and visit the site of his pre-coronation Court at Perth and of his coronation at Scoon. I arrived in Perth on a sunny evening in late May but I was disappointed to find that Gowrie House no longer existed. The grey stone quays of the once thriving port are still there, running alongside the wide, slow river Tay, but the House where Charles held his court has long since been demolished to be replaced with substantial houses of prosperous merchants which in their turn have become offices, as the maritime trade of the town has decreased.

It was early evening as I walked along the side of the river. The sun was low in the sky and the light was turning golden. All that remains of Gowrie House is a commemorative plaque, placed by the town council to show where the house and its riverside gardens had once stood. As I looked out across the river towards a sweep of hills which could have changed very little since Charles had sat at the point, I thought of him fretting under the increasing indignities which the Kirk kept forcing onto him. When he arrived at Perth he had just been told that he would have to sign a declaration which not only admitted his own current sins but which also repented for his father's sin in marrying into 'an idolatrous family'; for his own 'ill education and former wickedness'; and for having made agreements with the papist Irish. Charles had appealed to Argyll who had advised him he must make the public declaration, writing to him:

> *When you come into your kingdom, you may be more free, but for the present, it is necessary to please these madmen.*[51]

I thought of Charles reading that response, sitting looking at this same view; seeing the same light of the evening sun highlighting the distant hills and wondering if he would ever be crowned. The words Charles had written at that same spot three hundred and fifty years earlier came to my mind.

> *Nothing could have confirmed me more to the Church of England than the treachery and hypocrisy of the Kirk.*[52]

It was soon after writing this comment that Charles seemed to have cracked under the remorseless pressure of the Kirk. On the morning of 4 October 1650 he rode out of the back door of Gowrie House and set off towards Fife at a full gallop. Charles met up with Lauderdale and his advisors at the House of Lord Airlie. It was two days before Charles gave himself up to Argyll's Covenantors who were pursuing him. When he did so he was alone, except for his man servants. The Engagers having spoken and advised him had melted back into the mountains.[53]

The man who had set up this clandestine meeting is generally believed to have been Sir Alexander Frazier, who had been Charles's personal physician when he had first fled to Jersey and had also attended him in France. Frazier had later returned to Edinburgh and when Charles came to Scotland was again put in charge of the king's health by Argyll.[54] Frazier was a personal friend of Sir Robert Moray.[55] I was reminded of a comment that Stevenson had made about Sir Robert's attendance at the Lodge of Edinburgh in July 1647, he had said about Sir Robert:

> *The following year he was in Scotland, attending the Edinburgh Lodge on the occasion of the admission of one of the King's physicians.*[56]

Could that King's physician have been Sir Alexander Frazier, I wondered? Frazier was in Edinburgh at the time, he was a friend of Moray's and he could not have been attending the king for Charles I was the prisoner of Cromwell at Holmby. Had Moray proposed Frazier into Freemasonry, as well as into the Royal Society, I wondered?

Whatever the link, Charles had his meeting with the Engagers and returned to Perth prepared to please the madmen of the Kirk. From then on he would repeat what ever was asked of him, but in such a dull monotone that it was clear his words carried no conviction. He seems to have adopted the motto Sir Robert used himself 'to be, rather than to seem'. No matter what he repeated in public he could retain his own inward belief. His new tactics finally worked against the endless calls for greater and greater humiliation the Ministers of the Kirk required before they would crown him. He would repeat, parrot-like, any and every statement the ministers made. They could no longer argue or preach at him because he immediately agreed with them. His stance made the clergy look foolish and they started to lose face with the people. When Argyll told him he had behaved wickedly in proposing to admit malignants into the Army Charles calmly replied, 'We are all malignants to God.'[57]

The gathering public unrest with the tactics of the Covenantors forced Argyll to allow the Engagers back into the Scottish Parliament; to allow Lauderdale to meet with the king; and to allow non-Covenantors to serve in the Army. Charles II, King of Scots, Patron of Science and Pleaser of Madmen had been well advised by somebody!

Pleasing the Madmen

The following morning I rose early and drove the mile or so upriver towards the Palace of Scoon. Moot Hill is inside the grounds of Scoon Palace, which is now a popular tourist destination. By arriving as the gates were opened I hoped to get a few quiet moments inside the small chapel which stands on the coronation summit before the busloads of tourists started to arrive.

Romantic legend says that Moot Hill is formed from the earth carried on the boots of the nobles who have come over the ages to pay homage to the kings of Scots at this sacred place. The reality is that the mound is Neolithic, built by the long-gone megalithic society for its own purposes. Today, the artificial, flat-topped mound has a small mausoleum and a replica of the Stone of Scoon standing in front of it. The original is now kept at Edinburgh Castle along with the honours of Scotland, which were used at Charles's Coronation.

Scoon Palace was run down and almost in ruins when Charles was crowned there in 1651. The Abbey and Palace had been badly damaged by John Knox almost a hundred years earlier and the present Palace was not built until 1802. However, as I walked through the magnificent rooms of the modern house I was struck by the prolific use of the five-pointed star in the decoration

I should not really have been surprised to see these symbols because I already knew from Stevenson's work that Moray had been particularly impressed with this important Masonic symbol because it figured in his family's coat of arms. Sir Robert was descended from Freskin, Lord of Duffus in Moray, from whom his family took their name of Moray.[58] The lands and ruined Abbey of Scoon had been given to a branch of the Moray family in 1604 by James VI. Before that it had belonged to the Gowrie family, the builders of Gowrie House, who had been involved in a plot to kill James. Sir David Moray helped to save the king's life and as a reward was given the Gowrie lands of Scoon and the titles Lord Scoon and Viscount Stormount.

Little remains of the house Charles II visited apart from the Inner Hall, which still has the fireplace at which Charles would have been able to warm himself after the Coronation. And he would have needed to, as the ceremony took place on New

Year's Day, 1651, in the ruins of the old Abbey. The present Mansfield Mausoleum was built on those ruins in 1807, and contains many more five-pointed stars of the Murray [as they now spell their name] family crests.

I went out into the bright sunlight of the May morning and stood by the replica of the Stone of Scoon, in front of the entrance to the chapel on Moot Hill. The hill was deserted apart from two abandoned strimming machines, their operators evidently disappeared into the house for a morning cup of tea. As I looked down at the stone I thought of Charles coming here so long ago and the ancient prophecy of St Columba went through my mind. 'Except old seers do feign and wizard wits be blind. The Scots in place must reign, where they this stone do find.' I stood in the warmth of the morning sun remembering the events of that New Year's Day some three hundred and fifty years before.

Charles came by river from Gowrie House to the front of old Scoon house. Six nobles held a state canopy to cover him as he was escorted onto the ancient mound. The ceremony started with Charles publicly affirming his acceptance of the Covenant, standing on the very spot where I now stood, on the front of Moot Hill, before being escorted into the ruins of the abbey. Behind him came Argyll, carrying the crown of Robert the Bruce, which had also been worn by James VI.

The Moderator of the Kirk gave a long, remorselessly intense sermon on the subject of the sins of the Stuart family and the wonders of the Covenant. Apparently this sermon lasted for three hours.[59] When the sermon ended Charles publicly accepted the terms of the Covenant for the second time that day. He then took the coronation oath and was dressed in the robe of state and given the great sword of Scotland and the spurs of state were placed on his boots. Argyll now stood before him, holding the crown, while another lengthy sermon was given on the theme of preserving the crown from the sins of the predecessors of King Charles II. At last the crown was placed on Charles's head. But he still had to receive the sceptre. There was, of course, another sermon on the virtues of the Covenant, before he was finally allowed to hold it!

Now Charles, robed and with all the honours of Scotland bestowed on him, the sword, the spurs, the crown and the sceptre, was led to the crowd assembled around Moot Hill. There the new king stood while the *madmen* of the Kirk gave yet another sermon. Then as the final act of the coronation, John Middleton, a Royalist Follower of Montrose, returned to the cold fold of the Covenant, wearing sackcloth and ashes. This contrived scene of ritual humiliation, featuring the errant Middleton, made a strange climax for the ceremony, which had taken eight hours, most of it outside on a Scottish winter's day. Dawn had been breaking when Charles arrived at Scoon House river steps, the sun had set on the brief Northern day before his procession went inside the Inner Hall to eat the Coronation Banquet and he was allowed to warm himself by the old fireplace where I had stood a few minutes earlier. At least the Kirk fed their new king. The accounts of payments made for the feast show he had some of his favourite beef, along with ten calves heads, twenty-two salmon and many brace of partridges.[60]

The Moray family were formally represented at the coronation by David Murray [Moray], 4th Viscount Stormount and 4th Lord Scoon who took part in the ceremony.[61] I went back inside Scoon Palace to the Duke of Lennox's room, where the present Lord Mansfield has set up a display of exact replicas of the honours of Scotland and the Crown of the Bruce. I stood looking at the regalia Charles had worn before turning to gaze out of the window towards the ancient flat-topped mound of Moot Hill. Surely, Sir Robert would have been here for the coronation. It was a Scottish occasion he could not have allowed himself to miss. I knew it would have been physically possible for him to have attended from Edinburgh but I have not been able to find any confirmation. Just another romantic whim perhaps but plausible!

Once Charles became king he was kind to Sir Robert. He made him Lord Justice Clerk of Edinburgh, a strange choice for a man with no legal training, but perhaps loyalty mattered more than professional expertise.

Within eight months of his coronation the King of Scots was again a fugitive. Charles was defeated by Cromwell at the battle of Worcester. Lauderdale was imprisoned. Cromwell had taken Perth as well as Edinburgh while Charles disguised as a common labourer was fleeing on foot to France.

Moray remained in Edinburgh, although he was thrown out of his public office by Cromwell in April 1652. It was not a good time to lose his job, since his wife was both pregnant and mortally ill. Meanwhile, the sackcloth-wearing John Middleton, who had been one of Charles's senior officers at Worcester as well as the climax of the Kirks's entertainment at his coronation, had been captured by Cromwell. He escaped and fled to the far north of Scotland, where he led an uprising in mid-1652 which took Inverness and then petered out due to lack of support. Moray and his father-in-law, the Earl of Balcarres had set to work immediately after Sophia's death to try and organise an Army to retake Lowland Scotland for the exiled king. Balcarres intended to lead the movement himself but he was taken ill and Charles, working on limited intelligence from his base in France, appointed the Earl of Glencairn to take his place. Glencairn was a bad choice as he was disliked by Balcarres, Middleton and Moray. Balcarres and Moray wrote to the king urging him to replace Glencairn with Middleton. Glencairn responded by accusing Moray of plotting to assassinate the king and imprisoning him, with the intention that he should be put on trial for his life. Late in 1653 Moray wrote to Charles protesting his innocence.

I have already mentioned that one of the ways Masons can communicate privately while in public view is to use certain familiar phrases which outsiders do not recognise. In a personal letter to the king Moray protested his innocence but he did so using a form of words only a fellow Mason would recognise. Having assured the king that he was innocent (it was later proved that the incriminating letter cited against him was a forgery) he said:

> *And then having found me guiltless, your Majesty may, as a Master Builder doth with his materials, do what you wish with me.*[62]

Stevenson comments that Moray appears to be intending Charles to recognise the Masonic reference,[63] but this begs the question of how Charles would understand it. The reference is very carefully chosen for the Master Builder, to whom Moray refers, had suffered death rather than betray the sacred trust reposed upon him. Moray was a skilled politician and negotiator. He was arguing for his life. Why would he use anything less than the most powerful metaphor to convince his king? But metaphors only work when there is a shared symbolism. For Moray to expect Charles to be convinced by the words he chose, he had to believe he was speaking to a fellow Mason. I began to wonder if Charles had been made a Mason during his time in Scotland? I did not have enough evidence to decide so I put the question aside to consider later. Whatever, Charles must have been convinced by Moray's Masonic plead because he wrote to Glencairn to 'most carefully investigate the allegations' before taking any action against Moray.[64]

The Glencairn rising was a failure and by July 1654 Glencairn's troops were defeated at the battle of Loch Garry. Meanwhile Moray was freed and went to Paris, where the King was staying. He was still with the king's small court when it moved to Bruges in 1656.[65] At this time the king's relationship with the Dutch was getting very difficult. From July 1657 to September 1659 Moray lived in Maastricht, where he apparently used his Masonic contacts to collect as much information about the Dutch as he could. One motive he may have had could have been to try to get the Generality, which ruled the Netherlands, to prove more favourable towards Charles's cause. What is obvious is that as soon as it began to look likely that General Monck was about to act to restore Charles to the throne of England, Moray returned to the king, who was then in Paris. Moray remained in Paris, when, to make Monck's task easier, the king moved his court to Protestant Breda. Moray's long-standing links with Mazarin would have been useful to Charles at that difficult time, when Charles had to personally distance himself from his French family if he hoped to be restored to the throne of England.

When the king returned to England around the end of May 1660, Moray remained in Paris. He did not return to London until late August. But on his return the king was pleased to see him. Bishop Burnet, who witnessed the meeting, comments that:

> *His majesty received R. M. [Robert Moray] with much crushing and shaking of his hand, and with good looks and as much kindness as he could wish.*[66]

This was the period during which Charles established Moray in a grace and favour house in the Palace of Whitehall and also asked him to set up a laboratory within the Palace, where Moray might continue his studies of nature and science. Did this common interest in science come from a shared exposure to the ideas of Freemasonry?

Within twelve weeks of arriving in England Sir Robert was so well accepted by the philosophers who met at Gresham

College that they had chosen him to tell the king of their intentions! One can only admire the boundless enthusiasm of the man! In looking in more detail at Sir Robert's career I realised that his Scottish Masonic connections were of great importance in forming his ideas, informing his actions and providing him with a network of useful contacts.

Sir Robert Moray was slowly beginning to emerge from the mists of history. I remembered the words of Gilbert Burnet, Bishop of Salisbury:

> *While he [Sir Robert Moray] lived he was the life and soul of that body [The Royal Society].*[67]

Freemasonry had played a key role throughout Sir Robert Moray's career. It seems very unlikely he would suddenly abandon a lifelong interest, so just how much had his Freemasonry affected Moray's later actions in forming the Royal Society?

Conclusion

The more I discovered regarding Sir Robert Moray the more obvious it became that he had been a very important player in the Civil Wars. He had switched sides more than once as a young mercenary but he seemed to have formed a genuine affection for Charles II from about 1650 onwards. The importance of Freemasonry to Moray's thinking had been a continuous theme to which I kept returning. But the additional information I had collected had done nothing to help me understand how Moray had managed to get himself invited to the 28 November Meeting which Wilkins had chaired.

However, I was beginning to suspect that if I could understand Moray and the Freemasonry which had so influenced him, I would also understand his motives for founding the Royal Society. To continue my investigation I decided to try and find out more about Freemasonry in England during this period. Elias Ashmole, another early Fellow of the Royal Society had been at Oxford when Moray was knighted and was made a Mason in Warrington during 1646. Ashmole kept a quite complete diary so he looked like the next subject I should investigate if I wanted to understand the taciturn Sir Robert.

CHAPTER 7

Fellow Number Thirty-seven

12 Dec 1660: Mr Ashmole proposed Dec 12th 1660
26 Dec 1660: Mr Ashmole proposed

Entries from the MS copy of the Royal Society's First Journal Book

THE LIBRARY OF THE ROYAL SOCIETY stores some neat hand-written notes. They record what happened at its first meetings. These two entries are part of this journal. They show somebody was very keen to get Elias Ashmole into the group. He was proposed at the second meeting. Then, within a fortnight, he was put forward again! Had he been rejected in-between these two entries? There is no record of any such happening. Why was it so important to bring him into the new Society for Promoting Philosophical Knowledge by Experiments that he was proposed on two separate occasions?

It wasn't because he was a great scientist, as he contributed very little in the work of the Society, his only pretension to science was the study of astrology. But he did pay his fees very regularly and he was in favour with the king. There must have been many other possible candidates of whom this could be said, and they did not get proposed twice. Somebody 'really' seemed to want Ashmole and I was curious why.

Ashmole himself first mentions the Society in a diary entry dated 3 January 1661 when he says:

> *This afternoone I was voted into the Royall Society {blank} at Gresham College.*[1]

He left the blank because at that time, although the king had agreed to become the patron of the Society, he had not agreed its formal name. The word Royall was inserted above the line, probably at a later date.

On 3 January 1661, Ashmole added his signature to the extended list, now of 114, who had been proposed as fit and proper persons to be made members, at the meeting held on 5 December 1660.

Events must have been moving very quickly. Within five weeks of the very first meeting of the Society, Ashmole had not only been proposed twice, but had also been invited to Gresham College. There he was formally voted into the new Society as his diary records on 9 January 1661:

> *This afternoon I went to take my place among the royall Society {blank} at Gresham College.*[2]

After this entry he does not mention the Society again until the award of its First Charter. At that time he proposed to the King an extremely Masonic Coat of Arms for the new Society. His suggestion was not accepted. He did not seem to get to many meetings at Gresham, though by this time he had taken on the job of Windsor Herald. Perhaps he was too busy preparing for the coronation, which formed part of his official heraldic duties. However, the Society's records show that he paid his fees regularly even if he did not take an active part in the scientific work.

Historian Michael Hunter said of him:

> *Elias Ashmole was consistently regular in his payments, yet in all the information about Ashmole's life and activities there is hardly any evidence for his presence at any meeting of the Society after his admission, as there is hardly any in the minutes, but he was evidently a useful source of financial support though almost entirely inactive.*[3]

Sir Robert Moray died, suddenly, on 4 July 1673 and within a year the Society was having money problems. Ashmole was a poor attender, but he was one of the first fellows to sign a binding bond of payment to guarantee the Society's income.[4] The Society's minutes for 22 October 1673 listed fifty-seven persons 'that may be looked upon as good paymasters' and Ashmole's name is on this list.[5] On 7 March 1676 he gave an extra £5, above his normal subscription, to set up a fund to provide the Society with its own building.[6]

So why did Ashmole support the Society with his money, if not his presence, over so many years. I needed to know more about this man and what motivated him if I was ever to understand the Royal Society.

Master of Ordnance

Elias Ashmole was born in Lichfield on 23 May 1617, the only child of the local saddler. He went on to become both a solicitor and an astrologer. He is perhaps best known for the Ashmolian Museum. When he died, in 1692, he left his library and his collection of antiquities to Oxford University. This bequest started the museum.

Ashmole is well known to all Freemasons because there are many Masonic myths about him. For example a little leaflet,

Your Questions Answered (issued by the United Grand Lodge of England in 1999), proudly, but wrongly, claims, 'The earliest recorded making of a Freemason in England is that of Elias Ashmole in 1646.'[7] How and why Ashmole came to be 'made a Mason' at Warrington was something I did not fully understand. His motives, however, seem to have been far more political than many modern Masons imagine!

At the age of sixteen young Elias left Lichfield to move to London. He went to live with Baron James Pagit, his mother's cousin by marriage. At this time he began to keep an occasional diary of contemporary events. It is from this diary that I was able to discover so much about him. By 1638 he had qualified as a solicitor and married Elenor Manwaring, a young lady from Smallwood, near Warrington, whom he had met at the Pagit's house. She died, from the complications of pregnancy on 6 December 1641. She had gone to Smallwood to stay with her parents for the birth and Elias only learned of her death when he travelled up to Cheshire, intending to spend Christmas with her and his in-laws.[8]

In August the following year he wrote in his diary:

> *The troubles of London growing greate I resolved to leave the Citty & retyre into the Country.*[9]

He moved to Smallwood to live with his father-in-law, Peter Manwaring. During his stay in Cheshire he still worked as a solicitor. His diary tells us that in April 1643 he travelled down to London to assist Henry Manwaring, his late wife's cousin, on legal business. Ashmole shows his Royalist sympathies when he records that he did not approve of how:

> *divers Statutes & Pictures in the Abbey Church of Westminster were pulled downe & defaced by a committee of the House of Commons and members of the Trained Band.*[10]

On 27 May 1643 Ashmole travelled to Oxford. He was on legal business concerned with collecting the King's Excise from the town of Lichfield. Charles I, who had been driven out of London, had established his Court in Oxford. Ashmole decided to join Brasenose College and stay in the city. He wanted to study natural philosophy, mathematics, astronomy and astrology.

Ashmole's diary is unclear about exactly how he joined Brasenose. He took great care to erase the status under which the college accepted him. Certainly, there is no record of his graduating.[11] One almost suspects him of wishing to present the appearance of graduating without actually managing to achieve the reality. BA(failed) may be appropriate to describe his academic achievements. Indeed, he may simply have been living at the college as what is called a 'stranger'. Ashmole's uncle by marriage, Sir Henry Manwaring, was certainly a 'stranger' at Brasenose College at this time and he sponsored Ashmole to the college.[12]

While he was resident at Oxford, Ashmole started to pen parts of his diary in a cipher in order to conceal its contents from Parliamentary spies. He would continue this habit of secret writing throughout his life. The content of some of his number substitution codes (i.e. King, Queen, Prince of Wales, Col Bagot, Garrison, Soldier, Convoy and so on) show that he was well able to produce secret coded reports of a military nature.

On 22 March 1645 he met Captain Wharton, who was a senior officer in the King's garrison of Oxford and also a keen astrologer. Within a month Wharton had appointed Ashmole as one of the four Masters of Ordnance for the city. Towards the end of August, after his defeat at Naseby, Charles returned to Oxford. Ashmole notes that the king stayed from 28 to 30 August before leaving the city to its fate. By September Ashmole was working on the defence of the city against the expected Parliamentary attack. On 17 September 1645 he wrote in his diary:

> *This afternoone Sir John Heidon Leiftenant of the Ordnance began to exercise my Gunners in Maudelin Meadowes.*

This note was written in cipher so Ashmole was beginning to try and cover his back against possible denouncement to Parliament. By the end of 1645 Ashmole was Commissioner of Excise at Worcester, in addition to his military role. He kept the letter that he took to Worcester on 22 December 1645:

> *The bearer, one of the Gentlemen of the Ordnance to the Garrison of Oxford, having an Employment in your Lordship's Government, by the Parliament heere put upon him, Out of his desire to be made knowne and servicable to your Lordship has entreated my mediation and attestation to whose person, industry and merrits, during the tyme he hath been interested in his Majesties Service under my survey I can recommend him to your Lordships favour as an able, diligent and faithful man, wherein your Lordship may bee pleased to believe.*[13]

The letter was written to Lord Jacob Astley, Commander of the King's forces in the counties of Hereford, Worcester and Stafford. It was signed by Sir John Heidon, who had been so impressed when he exercised Ashmole's troops on Maudelin Meadows.

Ashmole arrived in Worcester, two days before Christmas. He was sworn in as Commissioner of Excise for the town on 27 December 1645. He lost no time in ingratiating himself with the local bigwigs, dining with Lords Brereton and Astley, to present his letter of recommendation. He proudly recorded in his diary that he had also met Sir Gilbert Gerard, the then Royalist Governor of Worcester.

During January, Ashmole was involved in helping Lord Astley prepare his forces to march to relieve Chester. Ashmole continually cast horoscopes trying to predict the course of the war and its likely consequences for himself. He was also interpreting his dreams to see if he could foretell the future. He recorded a dream in April that the king was marched out of Oxford and became worried about its meaning. On 27 April he recorded that he had dreamed:

> *The King went from Oxf: in disguise to ye Scotts.*[14]

This is one of the few accurate astrological predictions Ashmole made as Charles would leave Oxford and surrender to the Covenantors at Newark.

By that time Ashmole had become closely involved in the Royalist cause. He had bought himself a number of new and fashionable outfits to impress both his sponsors and various ladies he had met. Also, he had succeeded in persuading Lord Astley to transfer his military commission from Heidon's Oxford garrison to the one at Worcester. On 12 March 1646 he wrote in his diary:

> *I received my Commission for a Captainship in the Lord Astley's Regiment*

By 22 May 1646 Ashmole was appointed Master of Ordnance at Worcester and by 18 June he wrote in his diary that:

> *Col Washington met me at Seven Bridge and told me he was much beholding to me that I would take upon me this command and that I should do the King good service now he has so much waned.*

But he soon waned even further. Oxford fell on 20 June, leaving only Lichfield and Worcester holding out for the king.

Ashmole recorded the fall of Lichfield to Parliament on 14 July 1646 and ten days later he wrote:

> *Worcester was surrendered; and thence I rid out of Towne according to the Articles, and I went to my Father Manwarings in Cheshire.*

This is confirmed by the Calendar of State Papers where Captain Ashmole is described as being among the officers who surrendered. Worcester was the last garrisoned town to hold out for the King's cause and its fall signalled the end of the hopes of Charles I. As a Royalist officer Ashmole was forbidden to live within the bounds of the city of London and so was unable to earn his living by the practice of law. He went to the house of his dead wife's father in Cheshire because he really had nowhere else left to go. Ashmole had thrown in his lot with Charles I and by late 1646 he was out of a job and well out of political favour. Small wonder that he kept casting horoscopes asking if his luck would improve, could he safely travel to London and would he marry a wealthy widow?

Almost any wealthy widow would have done. Mrs Cole, Mrs Minshull, Mrs Ireland, Lady Thornborough, Mrs March, Lady Fitton or Lady Manwaring all figure in the erotic, and pecuniary, dreams that the 29-year-old Ashmole recorded during this period. His diary entries show him to be an inconsiderate and outrageous flirt as he tries to bring about a speedy marriage to help fill his 'leane Purse'. Typical of his entries are these examples of his dalliances with Lady Bridget Thornborough, Mrs Wall and finally Mrs March, none of whom would consent to marry him:

> *The Lady Thornborough sent for me and I went to her and found her in bed.*

*Mrs Thornborough lay upon a bed with me and exercised some love to me and that she did really love me. I felt her c*** and it seemed to be closed up.*

This night from 10 to almost 12; I discoursed with Mrs Wall in her bedchamber where still all her discourse beat upon her fear that she should not marry to please her friends.

[Writing of Mrs March who had consented to 'come undressed' to him while he was still in bed] I had this day divers kisses from her and she lent me her picture to wear next to my heart. She then put her hand into bed to me and protested she had never done so much since her husband died and she told me she hoped I was confident there was nothing she could afford me but I might command it.

Despite his romantic efforts Ashmole did not persuade any of these wealthy widows to marry him. He was forced to consider other means of making his living. The Articles of Surrender he had signed obliged him to either return to his home or 'go overseas and never to beare Armes any more against the Parliament of England or do anything wilfully to the prejudice of their affaires'. While he was trapped in besieged Worcester his mother had died; the wealthy Lady Thornborough had just refused his offer of marriage; and so his only remaining family was the father of his late wife, Mr Peter Manwaring of Smallwood, Cheshire.

For a while Ashmole scratched a living carrying out simple legal duties for his father-in-law but he seems to have been highly stressed. On 16 September 1646 he was feeling very sorry for himself. He began to develop boils on his arms and then things got worse. He confided to his diary, 'this night I first perceived a boyl to rise upon my a***'. His joints were aching and he became extremely constipated. Eventually he became so concerned that he began to record the number of stools he passed each day.

He continually cast new horoscopes asking if he should 'get a fortune by a wife without pains and easily'. Though what rich widow would want a constipated outcast with an enormous boil on his backside is a question he avoided asking his stars! He must have wondered about his attractiveness as a husband, however, as he also cast horoscopes asking if he would fare better if he went overseas. The young Ashmole was certainly ambitious and not shy of currying favour if he thought it would help his fortunes.

Then he suddenly seems to have resolved the question of how to proceed. On 17 October he borrowed some money from his cousin, Col Henry Manwaring, and bought a horse from Congleton Horse Fair. On 20 October he gathered his possessions together and by the 25 October was on his way to London, despite the undertaking he had given at Worcester not to live in the capital. He evidently believed that he would now be allowed to ignore the restraining agreement he had undertaken as Master of Ordnance at the surrender of the city.

His belief seems to have been well founded because by 20 November 1646 he was living in London and mixing freely with astrologers, alchemists and mathematicians. Among these was William Lilly. Lilly, I already knew, was a well-established astrologer, writer of a much-respected university textbook on Christian Astrology and a strong supporter of Parliament. Just what had happened to change Ashmole's fortunes and make him acceptable to a man like this?

The only material difference I could discover was that his cousin, Col Henry Manwaring, had introduced Ashmole to a lodge of Freemasons, meeting in Warrington. Ashmole had been made a Mason on the afternoon of 16 October 1646. His membership of the Craft seemed to be the key to meeting many influential people and allowing him to move to London despite the law. Indeed, a note in the papers of the Public Record Office, State Papers Domestic, Interregnum A. confirms the unlawful nature of his move to London when it says:

> *He (Ashmole) doth make his abode in London notwithstanding the Act of Parliament to the Contrary.*

Ashmole was changed almost overnight. He went from a despairing outcast, suffering from constipation, aching joints, repeated failures in love and with boils on his rear, to being an enthusiastic and bold adventurer. He was suddenly accepted in London society. The only change in his status was that he was now a Freemason. Had he become a Freemason to open doors for him at the highest levels in Parliamentary London?

Unfortunately, his diary entry only says where and with whom he became a Mason. It does not explain why he wanted to join the Craft. His diary entry for 16 October 1646 reads:

> *4H.30 P.M. I was made a free Mason at Warrington in Lancashire, with Coll: Henry Manwaring of Karincham in Cheshire.*
>
> *The names of those that were then of the Lodge, Mr: Rich Penket Warden, Mr: James Collier, Mr: Rich: Sankey, Henry Littler, John Ellam, Rich: Ellam & Hugh Brewer.*

The men making up the lodge are mainly local landowners, who would have been well known to the Manwaring family. Masonic historian Dudley Wright says of this meeting:

> *The proceedings at Warrington in 1646 establish some very important facts in the relation to the antiquity of Freemasonry and to its character as a speculative science . . . The term 'Warden' moreover, which follows the name of Rich Penket will of itself remove any lingering doubt whether the Warrington Lodge could boast a higher antiquity than the year 1646, since it points with the utmost clearness to the fact that an actual*

> *meeting of a subsisting branch of the Society of Freemasons was present at the meeting.*[15]

So here was evidence that Freemasonry was active not only in Scotland but also in England during the 1640s. Among the Masons present at Ashmole's initiation there are two of particular interest. The first was Richard Sankey, the father of the Freemason who transcribed the Sloane Manuscript. This document is one of the English copies of the 'Antient Charges', which had developed from the Schaw Statutes. It records the duties and privileges of a Freemason and gives an outline of what Freemasonry is about. Richard Sankey's son, Edward, dated his signed copy of the Antient Charges, October 1646. This was the very same month Ashmole was made a Mason.[16] So I now knew that the same philosophy of Freemasonry, which Robert Moray had learned in Scotland, had also been familiar to the Warden who initiated Elias Ashmole. It meant the Ashmole had probably been taken through a similar ceremony to the one Robert Moray had undertaken five years earlier in Newcastle.

The second interesting character at the Warrington Lodge was Hugh Brewer. He had been a Captain in the Royalist Army of James Stanley, Earl of Derby. Brewer had fought with Stanley against the Parliamentary forces of Lord Brereton to defend Warrington for Charles I.[17] He seems a strange choice to welcome Henry Manwaring, a serving Colonel of the Parliamentary Army into Freemasonry.

What is noticeable about the membership of this lodge is that it takes no view of the politics of its members. They are drawn from both sides of the Civil War. A Roundhead Colonel and two Royalist Captains as well as a number of local landowners from Warrington, Newton-le-Willows and Lymm, whose politics are not known.

As soon has he was safely in the Craft Ashmole became

revitalised. He stopped drifting and immediately started to prepare to move back to London. His astrological enquiries, as recorded in his diary, show he had been afraid to contemplate such a move before his initiation. What else had he learned from the Freemasons of Warrington? Perhaps a list of useful Parliamentary contacts in London and how to approach them?

Ashmole's biographer, C H Josten says of this time:

> *Perhaps his newly acquired Masonic connections had influenced Ashmole's decision. Certainly on his return to London, his circle of friends soon included many new acquaintances among astrologers, mathematicians, and physicians whose mystical leaning might have predisposed them to membership of speculative lodges.*[18]

This new circle of friends revolved around William Oughtred, the mathematician, alchemist, astrologer and inventor of the slide rule. Among Oughtred's friends were Seth Ward, Jonas Moore, Thomas Henshaw, Christopher Wren, William Lilly, George Wharton and Thomas Wharton. Also, within a year, Ashmole became a regular visitor to Gresham College, the institution which was so important as a meeting place for the founder members of the Royal Society.[19] By 17 June 1652 Ashmole was so well established in London that he was visited by two men I had already come across, John Wilkins and Christopher Wren. At this time Wilkins was a successful Parliamentarian academic. He was Warden of Wadham College and was reputed to be courting the sister of Oliver Cromwell.

Ashmole wrote in his dairy:

> *11H.A.M: Doctor Wilkins & Mr: Wren came to visit me at Blackfriers. this was the first tyme I saw the Doctor.*

Wren had just been appointed a fellow of All Souls, Cambridge.

Both Ashmole's visitors enjoyed the patronage of the Cromwell family. What had encouraged these two successful academics to visit this disgraced Royalist ex-officer, living illegally in London?

Did Ashmole's becoming a Freemason explain why they decided to visit him? Before I could be sure, I needed to see if I could find any more possible Masonic connections. I had been struck by the Masonspeak type comments in John Wallis's article entitled *A Defence of the Royal Society*. Wallis was a friend of Wilkins; an early member of the Royal Society; and also a friend of William Oughtred. I also knew that Oughtred was at the centre of a group of men, many of whom became founders of the Royal Society. He was never invited to become a member. Why was this? Quite simply because he died before the first meeting.

Oughtred helped Ashmole out by giving him lodging during the difficult times just before the Restoration. Exactly how Ashmole came to be first introduced to Oughtred is an interesting chain of events. Soon after arriving in London Ashmole wrote in his diary:

> *About a fortnight or 3 weekes after I came to London Mr: Jonas More, brought me aquatinted with Mr William Lilly. It was on a Friday night and I think the 20th: of Nov.*

Jonas Moore was, at the time, tutor in maths to the fourteen-year-old James, Duke of York. The Duke of York had been handed over to Parliament at the surrender of Oxford the previous year. Moore's appointment had been agreed by both Parliament and the king. William Lilly made a living writing almanacs of astrological forecasts, a sort of 'Mystic Meg' of his day. He wrote under the pen-name of 'the English Merlin'.[20] He was also a Royalist and a supporter of Charles I who tried not to quarrel with Parliament. It was Lilly who introduced

Ashmole to Oughtred and the circle of Parliamentary academics surrounding him.

In June 1647, only nine months after being made a Mason, Ashmole was asked by William Lilly to create an index for his text book, *Christian Astrology*, at the time a prestigious text widely used in universities for the teaching of astrology. True to form Ashmole cast a horoscope for the best time to start the work, and fixed on ten minutes after twelve noon on the fifth day of the month. The stars must have smiled on Ashmole because his association with Lilly increased his status in the 'scientific' society of London. Were Moore and Lilly Freemasons, whom Ashmole sought out soon after he joined the Craft? Had he been told by his new Masonic brothers in the North how to identify himself to his brethren in London? Unfortunately, he did not jot down his motives, in the way he did with his intentions of remarrying. Ashmole's *Diary*, however, is not a detailed daily record of his thoughts and life, such as the one Samuel Pepys left to posterity. Rather it is more a series of jottings about his business and military affairs. He discusses his attempts to marry and quotes a number of Masonic dedications, yet he only twice mentions going to Masonic meetings. Other contemporary writers have reported that he began to write a complete *History of Freemasonry*, where he may well have talked about his attitudes to the Craft but unfortunately this work has been lost.

Certainly, he was not afraid to advertise his new status in the Craft. He used blatant Masonic symbols on the frontispieces of his books. In addition he publicly accepted many Masonic dedications and tributes. All the evidence, including his own brief and casual note about attending a lodge in London many years later, indicates that he saw Freemasonry as a means to an end. Just as he cold-bloodedly set out to marry any available rich widow, so he seems to have intended to join a Society that would protect him if he returned to London; and which would provide him with a ready-made circle of useful contacts. This attitude seems to have worked to his advantage. What other reason, apart from Masonic preferment, could have persuaded the influential Oughtred to meet with this disgraced ex-officer? I made a mental note to keep a lookout for any additional information I could discover about Oughtred.

To see if Ashmole was using his Masonry to make useful personal contacts I decided to look more closely at his new friends. Exactly which people in Oughtred's circle might have been Freemasons? I decided to start with statements made by a senior member of that circle, mathematician John Wallis.

John Wallis, as I knew, was a friend of Seth Ward.[21] I was also aware that Wallis had written that the Royal Society had started with a series of meetings held at Gresham College 'about the year 1645'. These meetings he had described as being held under Masonic conditions, which forbade the discussion of religion and politics. Here I repeat his exact words:

> *Our business was (precluding matters of theology and state affairs) to discourse and consider of Philosophical Enquiries.*[22]

Among the many members Wallis mentioned as attending these meetings, he singled out the name of John Wilkins as important.

Ashmole arrived in London around the end of October 1646 and by 20 November he was welcomed into William Oughtred's circle, after being introduced to him by William Lilly whom Ashmole had met through Jonas Moore, yet another man who would become a founder of the Royal Society.[23] Within three months Ashmole was invited to attend a meeting of Mathematicians at Gresham College (on 16 February 1647) which John Wallis also attended.

I knew that Josten had said that the mystical leanings of

Ashmole's new friends 'may have predisposed them to membership of speculative lodges, yet it is not known of any of them that they belonged to the Craft'.[24] There is, however, evidence from Ashmole's diary that Wallis was friendly with a known Freemason. Ashmole says in his diary for 11 March 1682 that he 'received a Sumons to appeare at a lodge to be held the next day, at Masons Hall, London'.[25] Present at that lodge was William Hammond.

William Hammond was a Fellow of the Royal Society.[26] He had been proposed on 23 January 1661, within a month of the first meeting, and before Wallis himself, who was not proposed until 6 March 1661, although Wallis was elected earlier than Hammond.

William Hammond is described by historian Michael Hunter as a great lover of mathematics. He had translated many of Wallis's Latin works into English and so had worked closely with Wallis for many years.[27] I now knew from Ashmole's diary that Hammond had also been initiated into Freemasonry.

Here was written evidence that at least one of the group with whom Wallis was mixing between 1645 and 1648, the same group who had welcomed the disgraced Brother Ashmole in 1646, was a Freemason. Josten and other writers on Ashmole have speculated that most of this group were Freemasons. Their use of Masonic dedications to Ashmole support this view. Perhaps therefore it is not so surprising that they were following Masonic rules and adopting Masonic philosophy for the running of their meetings. Were those meetings really lodge meetings?

Professor Douglas Mackie said of Wallis's account of these early meetings:

> *In [Wallis's] the* Defence *of 1678 there is a detailed refutation of [the] . . . assertion that the Royal Society had originated in*

> *Wilkins's Lodgings in Oxford in 1659. Wallis stated that what he said was without disparagement to Wilkins, with whom he had been associated in the early meetings in London and in whose rooms in Oxford meetings were held, not in 1659, as Holder alleged but earlier, in 1648–9, although not earlier than the meetings in London, which continued there even after Wilkins, Wallis himself, and then Goddard, had in turn left for Oxford. Wallis was here clearly as concerned to give an earlier date to the Oxford meetings than Holder ascribed to them as he was determined to establish the occurrence of the still earlier meetings in London: and again he could speak as an eyewitness, because he was resident in Oxford from 1649.*[28]

Josten suspected that Ashmole's new friends were Freemasons. And now here was proof from Ashmole and the Royal Society's minutes that Josten was right in at least one case. If he was right about one, perhaps he was also correct about the others, although I knew it may never be possible to prove it.

However, all this group surrounding William Oughtred seemed to have been involved with Gresham College. Wallis, in particular, had gone to a great deal of trouble to record his view that the earliest meetings, which led to the formation of the Royal Society, had been held at Gresham College.

In 1940 Professor Francis Johnson, of Stanford University, put forward the idea that Gresham College itself was the inspiration for the Royal Society.[29] This idea has been difficult to sustain, as the organisation of Gresham College did not lend itself to the form of observational and experimental science that became the Royal Society. An additional problem for his idea is that when Wilkins, Ward, Goddard and Wallis all moved to Oxford, the meetings continued there, well away from any influence from Gresham College. However, if these gatherings were Masonic meetings, held at Gresham with the purpose of studying the hidden mysteries of nature and science, there is no problem in explaining how the meetings continued when the participants changed location. They would simply have formed a new lodge. In the light of this suspicion about Gresham College, and the people involved with it, I could now be fairly certain why Ashmole was invited into the new Society. It was not because of any pretension he had towards science but because he was a Freemason, with enough money to contribute to Bro Moray's project.

So the question I now needed to follow up was this: had Gresham College been a suitable venue for holding lodge meetings in the early 1640s? I decided to look more closely at the place where the founders of the Royal Society had chosen to hold their meetings.

The Role of Gresham College

Gresham College was established in 1579 as the result of a bequest in the will of Sir Thomas Gresham. In 1519 Sir Thomas was born into a family of merchants. He was particularly successful in his dealings and amassed a large fortune. Among his commercial achievements he built the very first paper manufacturing factory in England. He became the financial agent for Queen Elizabeth I and in this role set up and built the Royal Exchange. He had an abiding interest in education and in his will he left his house in Bishopsgate and an endowment, based on continuing revenues he received from the Royal Exchange, to found a college. It was to be named after him and housed in Bishopsgate. He left enough income to provide a home and a living for seven resident professors, each to be renowned in their own subject areas. The house also had meeting rooms to hold the regular public lectures that the professors were to give to the general public. Each lecture was to be given twice. Once in English for the benefit of the local population of London and then again in Latin so that foreign visitors might also benefit from the instruction.

Sir Thomas laid down in his will the subjects to be taught. They were: divinity, medicine, geometry, astronomy, rhetoric, law and music. From its earliest foundations, in 1598, the college encouraged its professors to discuss practical uses for their subjects. In particular the professors of geometry and astronomy worked closely with officers of the Royal Navy; with naval administrators; and with shipbuilders. Professor Johnson noticed these close links with the navy and saw that they all related to computational techniques for navigation and the efficient design of warships. He suggested that the practical bias of the early Gresham professors set down a basis for the practical science of the Royal Society. Professor Mackie agreed with him that Gresham's traditions were an important influence on the founders of the Royal Society. Mackie said:

> *Such historical facts as may be gathered about the Gresham College in late Elizabethan and early Stuart times suggest that it was the matrix in which the Royal Society originated.*[30]

No less than ten past holders of Gresham professorships became Fellows of the Royal Society when it got its first charter. These were: Christopher Wren, Walter Pope, Daniel Whistler, Laurence Rooke, Isaac Barrow, Robert Hooke, William Petty, Thomas Baynes, Jonathan Goddard and William Croome.

The Evidence of William Preston

It seemed prudent to check whether any Masonic literature mentioned Gresham College. I started by looking at the index of Gould's *History of Freemasonry*. There was no mention of either Sir Thomas or Gresham College. Next I looked through the detailed contents of William Preston's *Illustrations of Masonry*, published in 1772. I was delighted to see that there was a complete section devoted to Sir Thomas Gresham.

Preston claims that in 1567 Sir Thomas Gresham was appointed joint General Warden of Masons along with the Earl of Bedford. Gresham is described as 'an eminent merchant, distinguished by his abilities and great success in trade'. Preston goes on to explain that the reason two General Wardens were appointed was because while Freemasonry was well established in York the 'meetings in the South had begun to considerably increase'. The brethren in the North were overseen by the Duke of Bedford, on behalf of the General Assembly of Freemasons of York. Meanwhile Sir Thomas Gresham was put in charge of the Craft in the South. However, Freemasons in the South still had a right of appeal, 'on every important occasion' to the Lodge of York.

In the Appendix have discussed evidence from Edinburgh that in 1615 the Masons of York requested advice from the Lodge of Edinburgh on the detail of the ritual of the then main degrees of Freemasonry. This is said to be the formation of what is still known today as the York Rite of Freemasonry.[31] Preston says that this lodge at York dates from at least 1567 and seems to have the same rights over other lodges as Lodge Kilwinning successfully claimed over its adjacent lodges in the Second Schaw Statute of 1599. Did the Lodge of York have a similar, but undocumented role in England to that of Lodge Mother Kilwinning in Scotland I wondered?

Against the name of Sir Thomas Gresham, in Preston's text there was an asterisk. This referred to a detailed footnote which I have reproduced below:

> *Sir Thomas Gresham proposed to erect a building, at his own expense, in the city of London, for the service of commerce, if the citizens would purchase a proper plot for it. His proposal being accepted, and some house between Cornhill and Threadneedle Street, which had been purchased on that account, having been pulled down, on the 7th of June 1566, the foundation stone of*

the intended building was laid. The work was carried on with such expedition, that the whole was finished in November 1567. [There then follows a detailed description of the architecture which has been omitted] This edifice, on being first erected, was called simply the Bourse but on the 23rd of January 1570, the queen, [Elizabeth I] attended by a great number of her nobles, came from her palace of Somersett House in the Strand and passing through Threadneedle Street, dined with Sir Thomas at his house in Bishopsgate Street; and after dinner her majesty returned through Cornhill, entered the Bourse on the south side, and having viewed every part of the building, particularly the gallery which extend round the whole structure, and which was furnished with shops filled with all sorts of the finest wares in the city, she cauled the edifice to be proclaimed, in her presence, by a herald and trumpet 'The Royal Exchange;' and on this occasion, it is said, Sir Thomas appeared publicly in the character of Grand Warden.[32]

Preston goes on to add that:

During her reign, [Elizabeth I] lodges were held in different places in the kingdom, particularly in London, and its environs, where the brethren increased considerably, and several great works were carried out under the auspices of Sir Thomas Gresham, from whom the fraternity received every encouragement.[33]

William Preston wrote his *Illustrations of Masonry* in 1772. At the time there was a tremendous battle taking place between the Hanoverian and Jacobite traditions of Freemasonry. Preston belonged to the Jacobite tradition which accepted that Freemasonry had flourished under the Stuarts. The Hanoverian strand, whose beliefs still hold considerable sway in the present United Grand Lodge of England, preferred to believe that Freemasonry had been invented in London two years after the 1715 Jacobite rising. Historian of Freemasonry Dudley Wright sits firmly in the Hanoverian tradition, and when he revised Gould's *History of Freemasonry* he removed Gould's original opening section on Scottish Freemasonry and placed all Scottish references way back in volume iii. Wright would seem to be somewhat biased against any Stuart links with Freemasonry and so it is hardly surprising that when he revised Gould he tried hard to cast doubts on Preston's work.

Preston, however, was considered an important Masonic historian of his day and he became the Worshipful Master of the Lodge of Antiquity, which gave him the rank of First Master of the English Constitution. Preston says in the opening of his *Illustrations of Masonry*:

> *I was encouraged to examine with more attention the contents of our various lectures. The rude and imperfect state I found them, the variety of modes established in our lodges and the difficulties which I encountered rather discouraged my first attempt. Persevering in my design, I continued and assisted by a few friends, who had carefully preserved what ignorance and degeneracy had rejected as unintelligible and absurd, I diligently fought for and at length acquired, some antient and venerable landmarks of the Order.*[34]

Preston claimed Sir Thomas Gresham was an important Freemason. How accurate is Preston's claim likely to be? Sir Thomas had established Gresham College just before the reign of James VI(I), who I already knew to be a Freemason. Preston had nothing to gain by inventing a role for Sir Thomas Gresham but Wright would earn 'Masonic Brownie points' with English Grand Lodge by ignoring anything prior to 1717.

It occurred to me to check what Preston had to say about matters on which I was better informed, such as the role of

James VI(I) in English Freemasonry. Sure enough Preston claims that under James Freemasonry flourished in both England and Scotland and that 'lodges were convened under Royal patronage'. For the king to patronise a lodge he would need to be a mason, and as I have already mentioned, James had been made a mason at the Lodge of Scoon and Perth in 1601. Wright does not even mention James VI in five volumes. I needed to consider carefully what Preston put forward and then to see what other supporting evidence there was for his claims from other sources.

Preston records that James appointed the architect Inigo Jones as General Warden and surveyor to the crown. Jones was then 'deputised by his sovereign to preside over lodges'. He goes on to add:

> *Under his administration, several learned men were initiated into Masonry, and the society considerably increased in reputation and consequence. Lodges were constituted as seminaries of instruction in the sciences and the polite arts.*[35]

Preston goes further when he notes that in 1607 King James laid the foundation stone of a new banqueting hall at the Palace of Whitehall:

> *in the presence of Master Jones and his Wardens, William Herbert, Earl of Pembroke and Nicholas Stone Esq, Master-mason of England, who were attended by many brothers, clothed in form, and other eminent persons, invited for the occasion.*[36]

He adds that the ceremony was conducted with 'the greatest of pomp and splendour'.

Preston continues his story with the information that Inigo Jones remained in Masonic office until 1618, when he was succeeded by the Earl of Pembroke, 'under whose auspices many eminent, wealthy, and learned men were initiated, and the mysteries of the Order held in high estimation'. He adds that Jones continued to patronise lodges until his death in 1652.

About the Civil War Preston says:

> *The breaking out of the civil wars obstructed the progress of Masonry in England for some time. After the Restoration, however, it began to revive under the patronage of Charles II, who had been received into the Order during his exile. Some lodges in the reign of Charles II, were constituted by leave of several noble Wardens and many gentlemen and famous scholars requested at that time to be admitted of the fraternity.*[37]

A general assembly of masons, Preston says, was held at St Albans on 27 December 1663 at which Sir John Denham was made a General Warden of the Craft. Sir John Denham was one of the list of members proposed into the Royal Society at the second meeting and became Fellow number 42 of the Society. At the same assembly, Preston adds that Christopher Wren also became a Warden of the Craft.

In addition, Preston names Jonas Moore (F313) as a Freemason. I could not help remembering that Moore was the first new contact that Ashmole made within three weeks of arriving in London. He was also the contact who put Ashmole in touch with William Oughtred. By now I was fast becoming convinced that Josten had been right about Ashmole using his Freemasonry to better himself in London!

A final interesting comment Preston makes is that on 23 October 1667 Charles II worked a Masonic ceremony viz: 'the king arrayed as patron of the Craft levelled in due form the foundation stone of the new Royal Exchange'. There is an additional piece of evidence supporting this from the diary of Ashmole, where he records that he cast a horoscope for the most appropriate time for the stone to be set. As the stone was to be set in 'due Masonic form' Freemason and astrologer Ashmole would have been the best man to decide the best Masonic and Astrological time.[38]

It would appear Preston not only confirms many of the facts about Stuart Freemasonry but also adds detail to them. So perhaps I could take seriously the claim that Sir Thomas Gresham was a Freemason. But was there any separate evidence from early Masonic ritual that could show if Gresham College was inspired by the teachings of that ritual? It seemed sensible to try and discover the truth of the matter by consulting the earliest Masonic documents in England.

Some of the oldest written statements about Freemasonry are contained in papers which Freemasons call the Old or Antient Charges. These describe the history of the Craft of Freemasonry and the duties of Freemasons. The earliest Scottish copies of these Antient Charges are known respectively as the Kilwinning, Aitchison Haven and Aberdeen Charges. They all date from around the time of the Restoration, but in England there are some older versions. Apart from two very early manuscripts which mention Masons in passing, these documents, that list the principles of Freemasonry and duties of Freemasons, all seem to date from around the reign of James VI(I). This is the period I have already described during which William Schaw was creating the modern lodge system in Scotland. The oldest of these Old Charges, which lists suitable subjects for Masonic study and the duties of a Freemason, dates from 1583, i.e. four years after the establishment of Gresham College. It is known as 'The Grand Lodge Document'. But there are also about half a dozen broadly similar copies of these Old Charges written between 1583 and 1630. Two in particular stand out. The first is attributed to Inigo Jones who, Preston has said, was made General Warden of the Craft by James VI(I) and the other, as I knew, had been copied by one of the Warrington Masons who initiated Elias Ashmole.

The Inigo Jones document was bought by Masonic historian Rev A F A Woodford at public auction on 12 November 1879. He subsequently published the full contents of the document. The catalogue of the auctioneers, Puttick and Simpson, describes it as 'The antient Constitutions of the Free and Accepted Masons. A very curious folio manuscript, ornamented title and drawing by Inigo Jones, old red morocco, gilt leaves, dated 1607.' 1607 is of course the year that Preston reports JamesVI(I) to have carried out the Masonic Ceremony of placing the corner stone for his new Palace at Whitehall, attended in due Masonic form by the General Warden of the Craft, Inigo Jones, and his two Wardens, William Herbert, Earl of Pembroke and Nicholas Stone. An interesting coincidence of date!

As another interesting little aside I had also noticed that when Basing House, the garrisoned home of the Marquis of Winchester, had been sacked by Cromwell's Roundheads on 15 October 1645 they had taken Inigo Jones captive. Cromwell himself, reporting the victory to Parliament, said:

> *We have had little loss. Most of the enemy our men put to the sword, and some officers of quality, most of the rest we have prisoners.*

The Lord Protector, however, neither imprisoned nor killed Inigo Jones. Instead he instructed his troops to strip Jones naked and force him to stand and listen to a rabid public denunciation by Cromwell's personal chaplain, Hugh Peter. Then Inigo Jones was turned free to 'be taken away wrapped in a blanket'.[39] I have heard it said that Cromwell was also a Freemason. His lenient behaviour to an aged but senior Royalist is difficult to explain but would be straightforward if Inigo

Jones had been a Brother Mason. There is one unsupported but intriguing story which was popular among senior Freemasons early in the twentieth century which was published in the *London Daily Express* (29 January 1929) by W. Bro Vice-Admiral B M Chambers CBE, under the title 'Was Charles I Beheaded?' Worshipful Brother Chambers suggested in this article that Elias Ashmole took the place of Charles I on the scaffold and that the king lived on as Elias Ashmole. He based this strange theory on what he claimed as a well-known fact among Freemasons that: 'King Charles I, Grand master of the Freemasons, would not have been executed by men who were Freemasons themselves and they would have allowed another Brother to take the Grand Master's place.' Unlikely as the story is, it shows the strong belief among senior Masons that both Cromwell and Charles I were Freemasons.

The copy of the Old Charges, attributed to Inigo Jones, begins with a discussion of appropriate areas of study for a Freemason. Here is a transcription of that introduction:

> *THE ANTIENT Constitution Of the Free And Accepted MASONS 1607*
>
> *THE MIGHT of the FATHER of HEAVEN, and the Wisdom of the Glorious SON, through the Grace and Goodness of the HOLY GHOST, three persons and One GOD; BE with us and Give us Grace so to Govern us here in our Living, that we may come to his Bliss that never shall have Ending AMEN.*
>
> *GOOD BRETHREN and FELLOWS, Our Purpose is to tell you how and in what manner this Worthy Craft of MASONRY, was begun; And afterward; how it was kept and Encouraged by Worthy KINGS and Princes; and by many other Worthy Men.*

AND Also to those that be here; We will Charge by the Charges that belongeth to Every FREEMASON to keep; FOR in good Faith, If they take Good heed to it, it's worthy, to be well kept FOR MASONRY is a Worthy Craft, and a curious SCIENCE, and One of the LIBERAL Sciences.

THE Names of the Seven Liberal Sciences are these.

I. GRAMMAR, and that teacheth a Man to Speak and write truly.

II. RHETORICK, and that teacheth a Man to Speak fair, and in soft terms.

III. LOGICK, and that teacheth a Man to discern truth from falsehood.

IV. ARITHMETICK, which teacheth a Man to Reckon, and Count all manner of Numbers.

V. GEOMETRY, and that teacheth a Man the Mete and Measure of the Earth, and of all other things; which SCIENCE is Called MASONRY.

VI. MUSICK, which Gives a Man Skill of Singing, teaching him the ART of Composition; & playing upon Divers Instruments, as the ORGAN and HARP methodically.

VII. ASTRONOMY, which teacheth a Man to know the Course of Sun, Moon and Stars.

NOTE I pray you, that these Seven are contain'd under Geometry, for it teacheth Mett and Measure, Ponderation and Weight, for Everything in and upon the whole Earth for you

> *to know; That every Crafts man, works by Measure, He buys or sells, is by weight of Measure. Husband men, Navigators, Planters and all of them use GEOMETRY; for neither GRAMMAR, LOGICK nor any other of the said Sciences, can Subsist without GEOMETRY; ergo, most Worthy and Honourable.*

The Inigo Jones Document, which is one of the oldest English statements of the teachings of Freemasonry, reflects exactly the areas of study which Sir Thomas Gresham promoted when he set up Gresham College. The other documents from this period, the oldest dating from 1583, record the same main points. So Masonic literature of the late Elizabethan and early Stuart period records the attitude that all these seven liberal sciences, assuming that Sir Thomas interpreted Law as meaning the ability to tell truth from falsehood, are to be encouraged within the philosophy of Freemasonry. On this basis it would seem that Gresham College would have provided an ideal place to accommodate Masonic meetings. Its very foundation is based on the oldest recorded Masonic principles and its founder was said to be a Freemason.

Conclusion

Elias Ashmole had been an opportunist with an eye towards his own best interests. When he found himself on the losing side, at the end of the Civil War, he spent quite a while desperately trying to find a way to improve his position. Unable to find a wealthy widow to support him, he joined Freemasonry, with a view to using it to open doors for him in London. As soon as he was initiated he set off to London and immediately started to move in the Gresham College circles.

Looking more closely at Gresham College I found that in early Masonic literature its founder, Sir Thomas Gresham, was described as a senior Freemason. Gresham College itself was founded on the educational principles which were recorded in the Masonic documents of the period, known as the Old or Antient Charges.

The more I looked at the circumstances of the formation of Freemasonry the more it seemed to be that it and its philosophy of studying the hidden mysteries of nature and science had played a major part in moulding the attitudes of the men who made up the Royal Society. A phrase from an early Manual of Freemasonry, published soon after the formation of the Grand Lodge of London in 1717, echoed this idea. It said, 'The esoterical principle of Freemasonry is sun worship and science, as the basis of human culture and discipline.'[40]

Now I knew that Gresham College would have been an ideal place to hold Masonic meetings it was clear that I now needed to look more closely at the precise circumstances of the Restoration. What role had Charles II really played in the founding of the Royal Society, as king and possibly as initiated Freemason?

CHAPTER 8

The Restoration

The reign of Charles II cannot be understood by historians for whom English history is only the history of England. Lord Elton

What inspired him [Charles II]? . . . It was the vision of an England strong abroad and at home, her fleet triumphant, superior to the Dutch, supported and abetted by her natural friend France. Lady Antonia Fraser

THE SUMMER OF 1658 WAS STORMY, both meteorologically and politically. For the superstitious, who were the bulk of the population, there were many portents of doom. The summer was late arriving, with hail storms as icy as winter raging in June. A great whale, '58 feet long and 16 feet high, with a mouth so wide that diverse men might stand up in it', lost its way in the Channel and swam up the Thames to beach at Greenwich. The watermen of the Thames killed it with harpoons, eyewitnesses recording that while dying it gave out a horrid groan and a great spout of blood and water.

The military dictatorship of England was nearing crisis point. For nine years Oliver Cromwell had ruled, reviled by Royalists for killing the king and vilified by Republicans for betraying the revolution. The sheer force of his personality had compelled the warring factions in Parliament, and in the country at large, into a fragile peace. Now that stability and the small measure of religious tolerance he had introduced during his term of office were at risk. As the early hail storms of that last wild summer gave way to hurricanes in August the Lord Protector took to his bed with disturbing frequency. He was worried about his family. In July, Elizabeth Claypole, his daughter who had befriended Christopher Wren, was taken gravely ill. While rushing to visit her the coach in which he was riding was knocked to pieces as it overturned. He escaped, shaken, unharmed but seeing the mishap as a bad omen. He was worrying about God. 'It is a fearful thing to fall into the hands of the living God,' he confided in his chaplain. He was suffering from gout and the ague, but the quack remedies used by the doctors treating him seemed only to increase his suffering.

At dawn on Friday 3 September 1658, Oliver fell into a final coma and died in the warmth of that afternoon's sunshine. He left a squabbling and divided land. He had been fearful of his impending death. Cromwell feared the judgment of his God and he feared for the fate of the republic he had struggled to govern since the execution of Charles I. Oliver had spent his last days asking himself who would be strong enough to control the fractious squabbles of the Army and Parliament at his passing?[1]

While Oliver lay on his sick bed at Hampton Court, a mighty storm had come from the southwest destroying the orchards and woodlands of the diarist John Evelyn. This random natural event was to have an important influence on the science of forestry because it focused Evelyn's attention on practical methods of restoring and re-cultivating his trees. Evelyn wrote of it:

> *That tempestuous wind, which threw downe my greatest trees and did so much mischiefe all England over. It continued all*

> *night, till 3 afternoone next day, and was south west destroying all our winter fruit.*

He was, however, much more terse about the death of Cromwell, writing on 3 September: 'Died that arch rebel Oliver Cromwell, called Protector.' Evelyn was not alone in his distrust of Cromwell's Protectorate.

Deciding who would succeed him must have been a hard decision for Cromwell, for as often happens with a strong leader, he had created no obvious heir in his court. He had recorded his last testament on 31 October 1658, but he put off nominating his son, Richard Cromwell, as his successor until only hours before he died.

Cromwell considered no royal claim to the country, in law there was no king. Parliament had made sure Charles I would have no heir apparent. They had kept the condemned Charles waiting in an antechamber by the Whitehall scaffold from 10.00 am until 1.30 pm while they debated and passed an Act forbidding the proclamation of any 'succeeding monarch by inheritance'.

Oliver's choice of new Lord Protector was difficult. His caution in delaying his decision as long as possible was fully justified. Within nine months Richard Cromwell, his son, had resigned unable to carry the burden of office his dying father had reluctantly placed upon his shoulders.

The Council of State had accepted Richard as their new Lord Protector but even while Oliver was being buried in state in Westminster Abbey the Army chiefs were starting to discuss new cabals. The old malcontents of the Long Parliament were flooding back to London to pick over the political scraps; the disenfranchised Royalists were getting restless; and the organisation of the Fifth Monarchy, which were convinced Christ would arrive within days, had started to openly rouse their followers. The Fifth Monarchists were led by Thomas

Harrison, who had calculated that the second coming of Christ was due in 1660 and that He would establish a new Monarchy in that Fifth Millennium of the World. Harrison based this idea on the seventh Chapter of the Book of Daniel and the rather dubious arithmetic of Bishop Usher who claimed to have exactly dated creation. They were convinced that when Christ returned he would initiate a new rule of the Saints.

Richard Cromwell gave up the task of trying to create order within these opposing factions in June 1659. Under pressure from a popular hero of the Battle of Marston Moor, General John Lambert, Richard resigned as Lord Protector. Within a month the survivors of the Long Parliament, now known popularly as the Rump, had reformed themselves. They attempted to cashier Lambert, but he used the Army to suspend them. The apprentices of London formed a mob, called for immediate elections for 'a Free Parliament' and had to be subdued by force. Meanwhile, the army in England was itself fast falling into a state of disarray and Lambert's attempt to impose his own military dictatorship was not firmly based. He had no money to pay his foot soldiers and groups of them mutinied. On Boxing Day 1659, taking advantage of the disorder of Lambert's army, Speaker William Lenthall, 'with the Mace before and the Rump behind' processed through London to take possession of Westminster and re-establish what remained of the Long Parliament. The Government of England was fast crumbling into total anarchy.

The Commander in Chief of the army in Scotland, General George Monck now became involved in the confusion. 'Old George' as he was known to his soldiers, at least when he was out of earshot, was the first professional English soldier to believe that the military should be subordinate to the civil in matters of government.

Monck had started his military career as a commander in the

Royalist Army of Charles I until he was captured at Nantwich by Parliamentary forces in 1644. He was imprisoned in the tower of London until 1647, an event which persuaded him to switch sides. He did such a good job of convincing Parliament of his change of heart that he was made Governor of Ulster. When Cromwell invaded Scotland in 1650 he went along as Lieutenant-General, and on Cromwell's return, remained there as Commander in Chief. Monck also had some experience of Parliament and its ways, as he had taken part in Cromwell's ill-fated Parliamentary experiment.

Cromwell had tried to make an end of his military dictatorship in 1653. He was unhappy with his role and so had tried to create a nominated Parliament composed of wise and godly men who would honestly devote themselves to the task of remaking a shattered England. Perhaps he remembered Charles I's last words on the scaffold. 'What the country longs for is good government, not self-government.' To achieve this he asked the independent churches in each shire to nominate candidates who were 'persons fearing God and of approved fidelity and honesty'. It was in this ill-fated Barebone parliament that General George Monck had sat.

When Richard Cromwell resigned and the Rump Parliament tried to seize power, Monck became concerned about the problems of a Parliament without any checks or balances. He carefully secured his strongholds in Scotland; disarmed his extremist religious officers who were mainly the Anabaptists; and assembled his troops at Coldstream. He told them he was planning to march into England 'to assert the freedom and rights of three kingdoms from arbitrary and tyrannical usurpations'. Rumours spread that General Lambert was mustering an army in the South to march against him and yet another Civil War seemed about to start.

On New Year's Day 1660 Monck crossed into England. In Yorkshire he was joined by Thomas Fairfax on the long march south to London. Monck was not opposed as he marched London-ward but he was continually presented with petitions for a Free Parliament. The Rump Parliament distrusted him and tried to undermine him but on 3 February he entered London at the head of his troops. He arrested Lambert and sent him to the Tower.

Speaker Lenthall asked Monck to take the Oath of Abjuration of the House of Stuart, but Monck refused. The City Fathers of London had taken a decision to pay no tax until a Free Parliament was elected to replace the Rump and so Lenthall presented Monck with a serious dilemma. Acting on the authority of the Rump, Lenthall ordered Monck to march on the city and pull down its gates and portcullis. If Monck obeyed the order he would bring down the hatred of London on himself and if he refused it he would be compromising his own well-publicised position on military subservience to civil authority.

On Thursday 9 February Monck called a council of his senior officers and they composed a letter to the Rump. It proposed an issue of writs to allow the excluded members of the Long Parliament to take their seats and soon afterwards call a formal dissolution to make way for a Free Parliament. The letter was sent to Lenthall and on 11 February Monck, marching into the city at the head of his troops, announced to the Mayor and Aldermen, in front of the Guildhall, what he had done. The whole of London broke out into a wave of spontaneous celebration. Bonfires were lit and large 'Rumps' of beef were roasted over them. All the church bells rang out in celebration. Samuel Pepys confided to his diary that in many taverns and wine vaults the people were drinking the health of the exiled Charles II.

On 16 March 1660 the Long Parliament met for one last time, to vote itself out of existence. As it disbanded it called elections for a new Parliament. Meanwhile, Monck dispatched Sir John Grenville to Flanders to negotiate with Charles II about a possible restoration of the monarchy. In the mean time, General Lambert had escaped and formed an army in the Midlands. The elections resulted in a new Cavalier-dominated Parliament. Lambert, however, was totally opposed to the Restoration, and still posed a serious threat. Fortunately for the king, Monck defeated Lambert at Daventry and exiled him to the Channel Islands before the new Parliament met.

By the end of April the Cavalier Parliament had asked the king to return and rule them and they voted that the future constitution of England lay in the 'King, Lords and Commons'. The king was proclaimed in Westminster and part of the Fleet dispatched to The Hague to collect him and bring him to Dover. Fortunately, the diarist Samuel Pepys, in his capacity of Secretary to the Navy, was aboard the flagship *Naseby*, soon to be renamed the *Royal Charles* and he recorded many of the events.

A Deep Concern for the Navy

The abiding impression Charles II had of England when he returned in May 1660 was that of a country that had no sense of quiet. While his people celebrated his return he had to endure the unrelenting discharge of guns in a endless series of Royal salutes. Even as the noise of the guns died away, the continuous cheering of his newly cherished subjects left him in a semi-dazed condition. After the first week of this treatment he was still no further than Kent on his triumphal progress towards London. He received a tumultuous reception when he reached Canterbury. Samuel Pepys described it saying: 'the shouting and the joy expressed by all is past imagination'.[2]

That evening Charles wrote to his sister, Henrietta Anne: 'my head is so dreadfully stunned with the acclamations of the people that I know not whether I am writing sense or nonsense.'[3] Yet the following day 'Charles was fit enough to hold a Chapter of the Order of the Garter, in the damaged shell of the cathedral.'[4] He invested his old enemies, Monck and Montagu alongside his old supporters Southampton and Hertford.

Sir Edward Montagu couldn't be there in person, to receive his garter. He was still in Dover, aboard HMS *Royal Charles*, but his secretary, Samuel Pepys, recorded in his diary what happened early in the morning of Sunday 27 May 1660. Charles had dispatched his Herald, Sir Edward Walker, to meet with the Fleet. Sir Edward had served Charles since the death of his father, both during his time in Scotland and during his exile[5]. He was a skilful courtier who was well versed in ceremonial and making a grand impression. He arrived from Canterbury by coach soon after dawn. Pepys, as Montagu's Secretary, was dragged from his bed to receive the King's Herald. Pepys quickly issued instructions to the captains of the thirty ships of the fleet to assemble aboard the *Royal Charles* to witness the honour the king had done to their admiral. By implication, Charles was also honouring the men of Fleet as well. Sir Edward used his long experience of ceremonial to good effect. He made sure that Montagu, as he stood bareheaded on the windswept deck of his flagship before his assembled officers, was honoured in the company of every important naval officer.

The King's Herald carried the insignia and letter of appointment upon a crimson cushion. He stepped slowly and carefully through the ranks of officers, making sure he gave them all a chance to see and admire as he displayed the insignia in full splendour. The efficient Pepys had placed a gilt chair to hold the cushion and its contents. Sir Edward carefully placed the cushion down before reaching for the letter, sealed with the Royal Seal of King Charles II. He handed it to Montagu who broke the seal and opened the letter, before handing it back to the Herald to read aloud. The assembled company watched and listened as they and their commander were honoured:

> *To our trusty and well beloved Sir Edward Montagu, Knight, one of our Generals-at-sea, and our Companion-elect of our Noble Order of the Garter. The contents of the letter is to show that the Kings of England have for many years made use of this Honour as a special mark of favour to persons of good extraction and virtue (and that many Emperors, Kings and Princes of other countries have borne this honour) and that whereas my Lord is of a noble family and hath done the King such service by sea at this time as he hath done, he doth send him this George and Garter to wear as a Knight of the Order, with a dispensation for the other ceremony of the Habitt of the Order and other things till hereafter when it can be done.*[6]

Sir Edward then placed the be-ribboned George, the name for the insignia of the Order, around Montagu's neck and the garter on his left leg. He stood back and saluted Montagu as a Knight of the Garter. The watching sailors cheered their approval.

It was a smart move on Charles's part to honour Monck and Montagu; and he got double political mileage out of the gesture by investing Montagu in front of his junior officers. Investing Southampton and Hertford at the same time was also an inspired political move. Admiral Montagu had just escorted Charles on his voyage from The Hague. That voyage had only been possible because of the influence General Monck had with both the Army and with Parliament. Lord Southampton had been a senior advisor of the executed king. Charles had just recalled him from retirement to become Lord High Treasurer. Lord Hertford had fought for Charles I and had attended him up to the time of execution. By honouring these four old enemies together Charles made a strong gesture of reconciliation. He must have hoped this gesture would be copied by the other squabbling factions of his newly restored kingdom.

Showing a fine sense of drama Monck had gone to the port of Dover to be the first to receive the king on English soil. He had formally surrendered his sword, going down on his knees and making public obeisance to the young king. He then turned theatrically to the watching crowd and shouted 'God save the King!'[7] The crowd, of course, echoed the prayer and cheered the returning king. Charles in his turn embraced Monck and called him father. Arm-in-arm they had climbed into the same coach and set off together on the journey to Canterbury.

Monck, too, was an astute politician. To make sure that he stayed in the public eye, after humbly accepting his Garter, he honoured Charles in return. From Canterbury the king set off towards Rochester, abandoning the dubious delights of riding in an unsprung closed coach for the freedom offered on the back of a good horse. Likewise Monck was glad to get out of the carriage and get his feet back in stirrups. Together king and general rode to Blackheath where Monck had lined up his troops, the forerunners of the Coldstream Guards, for the king's inspection. As they rode along the lines of seasoned fighting men, Monck achieved two objectives: he reminded the king of the need to court the army's power; and that the Declaration of Breda had committed the crown to repaying the army's back pay, a matter the Rump Parliament had never seriously addressed.

Charles was playing the role of newly restored monarch with great skill. The ceremony of the Garter could easily have been carried out in London, but by taking time to do it in Canterbury he had achieved a number of objectives: he had consolidated his position with Monck and his troops and had won the support of the Fleet. Charles had also sent a message of reconciliation by sharing the honour with his late father's supporters. Beyond this the new king had sent a message of support to the Protestants of England, by conducting the Chapter in the seat of the Church of England whose head he was; and he had prolonged his journey long enough for him to arrive in London on the morning of his birthday.

King Charles II celebrated his thirtieth birthday riding into London. With his two brothers beside him, he was met by the Lord Mayor at Deptford and proceeded at the head of the great Royal Procession towards Westminster. He was accompanied by three hundred men, dressed in the finest doublets of cloth of silver; three hundred wearing velvet jackets; footmen in purple liveries; the Sheriff's men in their red cloaks, trimmed with silver lace; the London Companies, all dressed in their finest; and the aldermen of London in scarlet gowns trimmed with ermine. The Procession was finally escorted by General Monck, Lord Buckingham and Lord Cleveland riding behind a thousand buff-clad troops, whose uniforms had been specially trimmed with silver lace, and at the head of five regiments of horse. The King was returning in full glory.

The City of London was also dressed in its finest. Writing in his diary of that day John Evelyn said:

> *the ways were strewed with flowers, the bells ringing, the streets hung with tapestry, fountains running with wine; the Mayor, the Aldermen and all the City Companies in their liveries, chains of gold, banners, lords and bones, cloth of silver, gold and velvet. The windows all set with ladies, trumpets, music and people flocking the streets.*[8]

Imagine yourself standing alongside Evelyn, in the Strand, as the vast noisy procession flows by you. The sun is hot and strong on your face, and to obtain this position on the edge of the route you have been standing on that spot since the night before. Listen to the vast roar of the crowd as the king and his Royal brothers ride past you. See how the sunlight sparkles from their cloth of silver doublets and frames the mass of black curls of the king's hair. Smell the hot acrid sweat of the horses and the sweeter smell of fresh droppings. As the dust of the great procession slowly clears from your nostrils listen to Evelyn as his takes off his hat, lifts his face to heaven and thanks God for the Restoration of the king, overwhelmed that such a restoration has been brought about without the shedding of one drop of blood and done by the very army that rebelled against the monarchy. It will be another seven hours before all the people who have joined the procession since Rochester will finish streaming past the spot.

Maybe no drop of blood had been spilled during the Restoration, but Charles must have thought of blood on that day as he entered the Banqueting Hall at Westminster. Could he have avoided looking upwards as he entered the gateway, could he really have failed to look towards the spot where the scaffold had stood ready for his father, eleven long years before? The last words of his father must have sounded loud in his mind, haunting the loud loyal racket of his reception:

> *For the people I desire their liberty and freedom as much as anybody whomsoever; but I must tell you that this liberty and freedom consists in having government, those laws by which their lives and goods may be most their own.*[9]

When Charles caught sight of the place where his father had been executed he is said to have faltered. But he had to compose himself and listen, with apparent interest, to long speeches from both the Lords and Commons and then he in turn had to address the watching crowd of politicians saying:

> *The laws and liberties of my people, with the Protestant religion – next to my life and crown – I will preserve.*[10]

His comment gives an insight into the politics of this new king. He had learned self-control during his long exile; and he knew how to keep his own counsel, listen, agree; and to do exactly what he wished in the end. To rule he needed Parliament's support and he intended never again to lose his kingdom. He would perfect a technique of ruling that he had discovered in Scotland, and which would deceive everybody who came into contact with him. He appeared, even to his close associates, as a good-tempered, vague and lazy king, who left difficult decisions to his ministers and could be easily influenced by anybody who could contribute to his pleasures. As historian Hester Chapman comments:

> *Thus, when Charles prorogued his disobliging and rebellious Parliaments, the assumption was that he would not work against them during the intervals; for was he not seen doing everything but work – on the tennis-court, in his yacht, fishing, swimming or discussing some invention (new beehives for the Whitehall gardens, or the prevention of London smog) with his protégés?*[11]

From that first speech Charles laid down a pattern for a lifetime of successful political manoeuvres. He left the monarchy as secure as possible for his younger brother. This was the best he could hope to achieve and it really wasn't his fault that his brother James failed to capitalise on the opportunity Charles had given him. Charles, however, would probably have not been surprised by James's loss of the kingdom. Later in his reign, when James urged him to take greater care of his personal safety, he commented:

> *Brother, you may travel if you will – I am resolved to make myself easy for the rest of my life. I am sure no man in England will take away my life to make you king.*[12]

His reference to travelling referred to his years in exile, which he had no wish to repeat.

On the day of his formal Restoration Charles was intended

to go on to Westminster Abbey for a service of Thanksgiving but he declined, saying he was too tired. He prayed briefly in the Presence Chamber before retiring to eat his dinner in semi-public at the Palace of Whitehall. As he ate, the king is reported to have said to the crowd of onlookers surrounding him. 'I doubt it has been my own fault I have been absent for so long, for nobody that does not protest has ever wished for my return.' He paused for a moment before adding, 'Where are all my enemies?'[13]

Legends and rumours, probably encouraged by Charles himself, say that he spent the rest of that night in the bed of Barbara Palmer, née Villiers, a cousin of the Duke of Buckingham. Now at first glance this seems a rather romantic suggestion. He had spent the previous four days processing from Dover to London amid an enormous cacophony of joy and he had been almost overcome by the emotion of meeting his Parliament in the hall where his father had died. After fourteen hours travelling he had eaten his dinner in public, rather like an exotic animal at a zoo; and he still maintained the strength and reserves of charm to seduce the married cousin of one of his new allies. What kind of man's man was this newly returned king?

Yet the story is perhaps not quite as outrageous as it appears. There are a number of points to consider in its favour. Along with Charles himself, the Church of England was restored and with it came its full panoply of bishops and deans and they were keen to drive out the Puritans. It was these newly restored Bishops who were seeking to display Charles to their Puritan rivals. They intended to parade him at Westminster Abbey, in a service to mark their 'own' return to power. But why would Charles miss the public service to attempt to seduce a woman he had just met?

Firstly, Barbara's bed was not exactly virgin territory for Charles. She had already allowed him entry while he was still an exile. They had met in February 1660, when Palmer and her husband came to Brussels bringing a message from the Earl of Peterborough. Writers at the time say she was 'a sinister and exotic beauty'. Certainly, her portrait, painted by Lely in 1671, shows a voluptuous lady with jet black hair, a supercilious smile and a sexy glint in her eye. Charles gave away his intimacy with the lady when he casually admitted he knew she was immune to smallpox. His Chancellor, Edward Hyde was amazed the king had such knowledge of her, the sort of knowledge only her husband would have been expected to possess.[14]

Secondly, Barbara began to put on weight soon after the Restoration. She was safely delivered of her first daughter, Anne, almost exactly nine months after the king's birthday, on 25 February 1661. The king would eventually acknowledge five of Barbara's seven children to be his own bastards, and to give Charles his due they all received titles. Anne herself became Countess of Sussex.[15] If Anne was not the king's birthday present he must have provided the wherewithal for Barbara to conceive her very soon afterwards!

Thirdly, the direction of Charles's politics had been laid down in his response to the Loyal address of the Lords. He had already realised that the people of England would feel less threatened by a king who would rather take a beautiful woman to bed than to force himself through a religious ceremony. He must also have been aware that going to the service would have endorsed the Bishops of the Established Church in their attempt to gloat over the discomfiture of the various other religious groups who made up the bulk of his kingdom. On the day of his first reception in London he had been made aware that he owed the bloodless return to the support of the Presbyterians. He also knew, from bitter Scottish experience, they would not take kindly to him leaning too far towards the bishops. His life would be far simpler if questions about the Church were not aired too soon, so he ducked away from that issue and into Barbara's amorous embrace. And he probably enjoyed it more than the service!

His open seduction of Barbara, and the rumours he allowed to circulate about it after the event, can be viewed as yet another conscious act of conciliation between the warring factions he had to rule. His indifference to all forms of religion helped calm the national suspicions of the people that he might be unduly influenced by Catholic France. He realised that the general hatred of Popery at the time was a political, not a spiritual matter. He said of England, soon after his restoration, 'Beyond the sea it seemed as if people worshipped God in earnest, but here, in jest.'[16] His attitude to religion can best be summed up with the story told of one of his visits to church. The minister, in the midst of a long sermon, had to stop to berate one of the king's party with the words 'Hush Sir, you snore so loudly you will wake the king!' In 1663 Barbara converted to Catholicism[17] and Charles was asked what he intended to do about it. He replied that he never meddled with the *souls* of ladies pointedly leaving open to himself unlimited opportunities to meddle with their anatomy.

He also maintained a sense of humour regarding religion. When he gave an audience to Quaker leader William Penn, whose religious belief insisted he keep his hat on in the Royal presence, Charles removed his own hat. Penn asked why he did so and he is said to have replied, 'it is the custom in this place for only one person at a time to remain covered'.[18]

In the Declaration of Breda Charles had promised to deliver:

> *a liberty to tender consciences. No man is to be disquieted or called in question for differences of religion, so long as these differences do not threaten the peace of the kingdom.*[19]

Charles did try to keep this promise. By 25 October 1660 he had put together and issued a declaration to modify the

Episcopacy and he believed the Presbyterians would accept it. But, early in November, it was rejected by the Commons. Charles spent much of late November and early December setting up a conference at the house of the Bishop of London to try to work out some other compromise acceptable to the Presbyterians.[20]

He also had many other matters to contend with during the first few months of his return. The Chancellor, Edward Hyde, now Baron Hyde and soon to become the Earl of Clarendon was flooded with loyal petitions for preferment in the Church; in the peerage; or simply for financial aid. An example of the sort of matters Hyde had to deal with was a bill from the drapers of Worcester who had been ordered by Charles I to clothe his life guards in red jackets but had never been paid. They demanded £450.[21] These were not the only tasks the new administration had to address. The existing Parliament was not technically legal and a fully legitimate one had to be elected before a coronation could be arranged. Charles was faced with a multiplicity of decisions, all requiring his attention during the first twelve months of his restored monarchy and he had few supporters with a wide enough vision of the future to support him. Edward Hyde, his Lord Chancellor during exile, was proving to be too inflexible after the return to power. He never seems to have really understood Charles's technique of constructive laziness. Hyde became increasingly overbearing in his demands of the king. He is reported once to have told the king, 'Matters are in ill state but there is a good able man if Your Majesty would employ, all would soon be mended.' When the king inquired who, Hyde replied, 'This is one Charles Stuart, who now spends his time f*****g about the Court, but if you were to give him employment, he would be the fittest man.'[22] The pressure of his Chancellor, the demands of his petitioners, the quarrelling of the religious factions and the sexual demands of his mistresses must have left him with little spare time in those first hectic months of his return, and yet the record shows he made enough time to play a major role in setting up the Royal Society.

Why did it matter to him so? To answer that question I decided that I needed to understand the situation he inherited from Oliver Cromwell.

The Price of Cromwell's Dutch Treat

Now that he was king, Charles found that he had been left with a difficult relationship towards the Dutch. Soon after the Battle of Worcester, Cromwell had begun a naval war with them. Winston Churchill described it as 'the first war in English history which was fought for primarily economic reasons'.[23]

The Dutch had not been pleased when Cromwell had Charles I beheaded, but as fellow Protestants, in a very Catholic world, they felt obliged to try and come to some sort of deal with England. Cromwell had sent a mission to The Hague in 1651 to try to work out an agreement but the talks failed. He then introduced a Navigation Act that forbade the use of any foreign shipping for importing goods to England, except the ships of the country doing the exporting. This Act crippled Dutch trade with England and was one of the causes of the First Dutch War. There was already bad feeling between the Dutch and English East India Companies and this added to the general hostility. By the middle of 1652 Cromwell's fleet had caused considerable damage to Dutch shipping. This damage greatly reduced the ability of the Dutch to carry out overseas trade. They sued for peace only to be lectured by Cromwell, who said to them, 'You have appealed to the judgement of Heaven. The Lord has declared against you.' Cromwell's peace terms, however, were so harsh they were not even acceptable to his own Parliament. The members thought them too hard to inflict on a fellow Protestant country and it was not until Cromwell became dictator (Lord Protector) that he was able to force the terms on the Dutch. His peace terms against Dutch trade were so harsh, however, that he merely succeeded in sowing the seeds of a future war. These seeds were just beginning to sprout when Charles was restored to his kingdom.

After their defeat in the first war the Dutch had developed fast, light manoeuvrable ships and they now had a stranglehold on the Baltic trade; the spice trade with the Indies; and also dominated the herring fishing industry. The Dutch had ambitions to recover the colonies they had lost in the First War, both in the Indies and in America. What is more they still smarted from the ignominious terms of their defeat by Cromwell. On his restoration Charles accepted the congratulations of Johann De Witt, the Dutch Minister of State and suggested that he was interested in trying to work with him.[24] Charles would have been aware of a general dislike of the Dutch within England. Since the First Dutch War the Dutch had become better traders than the English and they had faster ships that cost less to run. In the Baltic, the West of Africa and in America the Dutch were strong trading rivals to the English. If they couldn't be defeated by commercial means, people were beginning to suggest that perhaps force should be tried. The Dutch fleet, however, was strong overseas and Charles must have seen it as a threat to the shipping of his country.

There was also another factor at work. Charles had a natural inclination to support his French cousin and Louis XIV had aspirations to invade the Spanish Netherlands. Charles may well have been worried that Louis would form an alliance with the Dutch, against the English, to make it easier to take control of the Netherlands. Louis was just about to marry Maria Teresa, the daughter of Philip IV of Spain. Philip's heir Carlos was sickly and congenitally retarded. When he died Louis would have a claim on the Netherlands. However, Charles was now in a position to receive up-to-date information about the intentions and attitudes of the Dutch. Sir Robert Moray had spent three years in Maastricht, from 1657 to 1660. He had returned to Paris early in 1660 but it is interesting to note that in March 1659 (soon after Cromwell's death, when a restoration of Charles was starting to look possible) Moray was presented to the town authorities of Maastricht by Everard, the master of the Craft of Masons of the town. He had taken the local Mason's Oath from Everard and been made a member of the Craft of Maastricht. Because of this link he was also made a Citizen of Maastricht.[25] His letters do not reveal the reason for this action, but on a number of previous occasions he acted as a spy. Was he collecting first-hand evidence about the intentions of the Dutch on behalf of Charles? Was it a coincidence that Maastricht, a strong Dutch Fortress garrisoned by about 5,000 men was the first target that the French chose to attack when Louis of France did invade the Netherlands? Sir Robert was an expert on fortifications from his days with the Covenantors.

The Navy Charles inherited from Cromwell had been weakened during the battles of the First Dutch War and since that time the Dutch had built up a stronger fleet. If there was to be a war with the Dutch then something needed to be done to strengthen the Navy, but Charles did not have the funds to do it. This is where the genius of Sir Robert Moray came up with a solution.

Moray had been in Edinburgh during the First Dutch War and a lot of the action had taken place off the coast of Scotland. Historian Lord Elton, says of the First Dutch War:

> *The Dutch lived by trade and their trade routes, to western Europe, the Mediterranean and the ocean, all passed close to the southern and eastern shores of their enemy. The English were still predominantly agricultural, and able, if need be, to dispense with trade and manufacture. They could therefore*

Galileo admitting the heretical depravity of believing the earth revolves around the sun, before the Most Eminent and Reverend Lord Cardinals Inquisitors-General. (*The Art Archive/Private Collection/Eileen Tweedy*)

The Right Revd Dr John Wilkins, FRS, Bishop of Chester. (*The Royal Society*)

The Honourable Robert Boyle, FRS. (*The National Portrait Gallery, London*)

Gresham College, founded by Sir Thomas Gresham, where the first meeting of the Royal Society was held. (*Mary Evans Picture Library*)

Wadham College, Oxford, where Dr John Wilkins was Warden 1648-59. (*Society of Antiquaries of London*)

Guy Fawkes and the Gunpowder Plotters who tried to assassinate James VI(I) in 1606. (*The Art Archive*)

Rosslyn Chapel – built in the fifteenth century by William St Clair, the founder of Freemasonry. (*A.F. Kersting*)

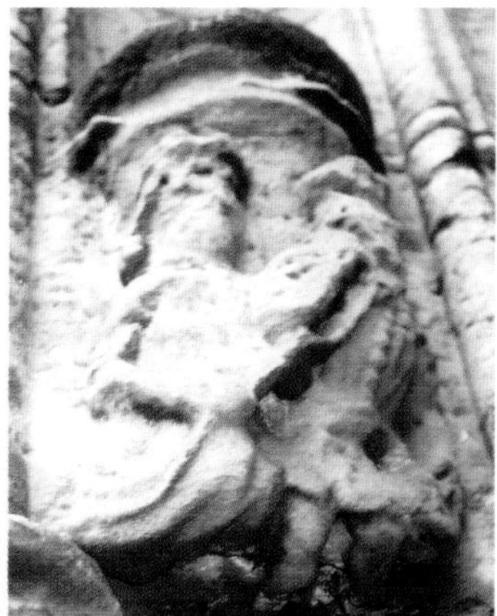

The 1460 statue of a Masonic Candidate carved in the wall of Rosslyn Chapel, showing the interesting arrangement of the Candidate's legs. (*G.R.J. Lomas*)

James VI of Scotland and I of England.
(*The National Portrait Gallery, London*)

Henry Oldenburg, one of the first Secretaries of the Royal Society and founder editor of the Philosophical Transactions. (*The Royal Society*)

James, Duke of York facing the Dutch Fleet at the Battle of Lowestoft on 3 June 1665.
(*Courtesy of Simon Gillespie Studio, London*)

The Frontispiece of the Royal Society's presentation copy of Sprat's *The History of the Royal Society*, endorsed by John Wilkins. (*The Royal Society*)

Oliver Cromwell, portrayed in a contemporary etching as a Pillar of the State. (*The National Portrait Gallery, London*)

The rural site of Greenwich, chosen as the site for the Royal Observatory. (*Museum of*

The observing room at the Royal Observatory, Greenwich. (*Mary Evans Picture Library*)

London/The Bridgeman Art Library)

Charles II, *Fundator et Patronus* of the Royal Society. (*The Royal Society*)

Nell Gwyn – one of Charles II's numerous mistresses. (*The Art Archive/Army and Navy Club/Eileen Tweedy*)

The Great Fire of London – September 1666. (*The Art Archive/London Museum/Eileen Tweedy*)

concentrate upon attacking Dutch trade, without exposing their own, so that the war developed into a series of escort actions in which the Dutch admirals would seek to pass great convoys, down the Channel or round the north of Scotland, where the English soldier admirals, Blake, Monck and Deane, barred the way . . . By the spring of 1654 they [the Dutch] were ready for peace.[26]

Sir Robert could not have avoided hearing about the convoy actions of the Roundhead Navy off the coasts of Scotland. When he was exiled in the Netherlands, after 1656, he would have been able to witness at first hand the effects of the Roundhead naval actions on the Dutch. This experience could only have reinforced his soldier's viewpoint that naval strength was important in any future battle with the Dutch.

The First Dutch War had really been about the beginnings of the British Empire. Spain had been the first country to exploit the Americas but by the mid-seventeenth century Spain was a spent force. The Dutch and the English had colonies in North America and Cromwell had been concerned about controlling them. By 1650, following the execution of Charles I, all England's colonies were in open revolt. Lord Elton says of their actions:

Although the rebels mostly made a somewhat unconvincing profession of royalism the various risings were in fact directed not so much against the victorious regicides as against English suzerainty itself . . . They flew the Royal standard, but nine out of ten of them felt much more strongly about the sugar, or tobacco, trade than about the wrongs of His Majesty. Indeed if the king had won the civil war he would probably have had to suppress a colonial revolt himself and for the same reasons.[27]

Oliver Cromwell sent a naval expedition against these revolting colonists in 1651. By 1652 he had succeeded in completely quashing them under the threat of naval bombardment. It is likely that this lesson would also have been learned by Moray, since he had met the son of one of the founding fathers of Massachusetts while attending Charles I at Oxford in 1643. This man was John Winthrop junior.

His father, John Winthrop senior, had been the first Governor of Massachusetts in 1629. Massachusetts began a trend for self-rule among the new American colonies. The colony had been founded by the Massachusetts Plantation Company, under a patent issued by the English Crown but this meant that the lines of command, between the Company offices in London and the settlement on the East seaboard of America, were long and tedious. In England that year there was considerable unrest in the Puritan communities. Parliament had been dissolved and Puritan leaders were being arrested and hounded by Charles I. Lord Elton comments that:

> *In many prosperous and influential English households serious men and women were beginning reluctantly and anxiously to consider fleeing from the wrath to come by transporting themselves and their children to the New World. Among them were several members of the new Company. To some of these a novel and revolutionary idea presented itself. Why not transfer the whole Company to New England, government, charter and all? Such a thing had never been done before, but in these days of crisis why stand upon precedent?*[28]

During the period of the Protectorate the colonies had prospered under benign neglect from Cromwell's England. They were, however, under threat from the increasingly confident Dutch by 1660. It is surprising that they had survived at all under Charles's grandfather James VI(I). His navy had become very run-down and would not have been able to protect the young colonies from a determined naval action by any of the European powers. The welfare and payment of his seamen had been neglected and the management of his naval dockyards had been a byword for corruption. Fortunately, during this period of establishment possible European predators had been distracted by the Thirty Years War (1618–48). By the time they were again looking westwards to the New World Cromwell had partly rebuilt the navy, under Admiral Robert Blake. Cromwell was able to use Blake to subdue the colonies and drive off the Dutch but he was no empire builder. His iron Puritanism and reluctant dictatorship did not encourage new colonies. He did, however, leave a small but respectable navy when he died. But Charles must have known that his fleet was nowhere near strong enough to protect his newly restored overseas assets from any determined attack by the Dutch and it had suffered a further two-year deterioration in the confusion since Cromwell's death.

Charles II inherited two major naval problems. Cromwell's reign had caused an unnatural repression in trade and overseas adventures, but with the Restoration there was a sudden surge of energy and expansionism. Along with this new energy came a greater interest in enlarging the existing colonies and founding new ones in the Americas. The First Dutch War had, however, failed to solve the problems of Dutch and English competition in America. New Amsterdam, to the north of New England, was thriving, and its Dutch masters also had expansionist plans. They had a large merchant navy and plans to make it even larger. The colonies were still bridling under the trade restrictions Cromwell had forced on them with his Navigation Act of 1651, which forbade the carrying of English merchandise on Dutch ships. The colonists of the New World were keen to trade with the Dutch and to cut out the run-down and expensive English Merchant Fleet.

His second problem was the French. There was a very real

danger of France joining in the game of empire and supporting the Dutch in America. One of Charles's first acts as king was to reinforce and extend Cromwell's Navigation Act of 1651 but legislation is worthless without the military strength to back it up. Charles simply did not have a strong enough navy. He clearly knew what had to be done. The Dutch were already a threat to the control and growth of the colonies and seemed to be developing into a threat to the security of his newly restored kingdom. But where was he to get the money and the expertise to develop his navy?

Charles needed more ships and the ability to sail them quickly to trouble-spots on the other side of the world. In those days ships were made of oak and to increase the production of ships meant being able to provide copious supplies of suitable timber. During the Civil Wars few land-owners had been able to devote themselves to the management of their estates. As a result good timber was in short supply. Even if he had his ships the secondary problem of navigating them to the right place quickly and safely was a difficult task. Latitude was a simple matter to determine but the measurement of longitude was the most awkward problem of navigation faced by any captain.

Charles needed a research team to study these problems but he could not afford to set up such a group. This is where Sir Robert Moray came up with a brilliant idea! From his correspondence with various scientists during the period of his exile in Europe he knew that some of the best scientists in England had not been in the Royalist camp during the rule of Cromwell. Could he bring together the best minds in the country and unite them in the task of rebuilding and strengthening the navy? Moray had no money to fund this venture and Charles had not been given anything like enough funds from Parliament to provide the capital, so if it was to be done it would have to be self-financing.

There was, however, a secret factor working in Moray's

favour. He knew, from his own experience in the Craft, that Freemasonic lodges encouraged the study of experimental science while at the same time forbidding any distracting discussion of religions or politics. When he visited London as a French agent, he came into contact with the group surrounding William Oughtred, who were associated with many recorded Freemasons. John Wallis was a leading member of the Oughtred circle, and he wrote that the first meetings of the men who were to go on to found the Royal Society met under Masonic conditions (i.e. the discussion of religion and politics was forbidden at those meetings). These meetings could have been lodge meetings, using their after-proceedings to extend the subjects of discussion. Many modern lodges have lectures and demonstrations following the formal proceedings. If this was so, it gave Sir Robert the connections to set up a scientific society able to support the king's desire to create a strong and effective navy. But there remains an outstanding question. Why should a group of disenfranchised and dispossessed Parliamentary scientists support the newly returned king? After all, one of the most influential of these men, John Wilkins, had just suffered greatly from Charles's ill will.

Historian Margery Purver commented on this situation saying:

> *Wilkins, formally, could do little. At the Restoration he had lost his Mastership of Trinity College, Cambridge, to which in 1659, he had been appointed from the Wardenship of Wadham College, Oxford; and although personally well liked by his colleagues in the Society, he was certainly not in a position to represent them in an approach to the King.*[29]

Wilkins could not approach the king but Moray could. Within a week of the very first meeting he brought back Royal words of support. The Society's own journal book records that on

5 December 1660 he reported that he had told the king about the new Society and the king approved of it.

When Sir Robert attended the meeting at Gresham College did he set out to encourage those Freemasons present to band together and form a new society? His suggestion was for a society that would advance their shared Masonic ideals in an atmosphere which would encourage experimental science so that improvements could be made to the state of the country's defences. As Wren and Boyle were certainly not Freemasons at that time, perhaps this is the reason why they were not included in the discussion and had to be asked to join at the second meeting.

An additional factor that Moray could have used to encourage the ill-used Parliamentarians into such a venture would have been to reveal that the king was also a Brother Mason. I have already mentioned that Charles's grandfather James VI(I) had been a Freemason so there was a precedence for the King of Scots also embracing the Craft. But, unlike the case of James VI(I) there is no lodge record of an initiation for Charles. What other evidence was there to support this idea?

There is William Preston's statement that Charles II had become a Freemason while he was exiled from England. How likely is this to be true?

As Charles fled to Jersey when he was only a teenager he would have been too young to be admitted to the Craft before his father's execution. The first opportunity would have been during his stay in Scotland. His grandfather had been initiated into the Lodge of Scoon and Perth so there is a possibility he joined while staying in Gowrie House. However, there is no record of this happening and the Lodge of Scoon and Perth has never claimed it. It is worth remembering that the Initiation of James into the Lodge of Scoon and Perth had resulted in considerable ill-feeling and the eventual rejection of James as Grand Master.

Many of Charles I's Scottish Court had been members of the Lodge of Edinburgh and this group later formed the Scottish faction known as the Engagers. There is a period of a few days, when Charles fled from Gowrie House to meet up with the Engagers, during which time his activities are not known. What is understood is that he returned to Gowrie House with a completely new strategy. 'To do everything asked of him but keep his own counsel.' It would have been completely in keeping with this new attitude for him to have become a Freemason at the hands of his Engager supporters in order to further consolidate their support for him. Certainly, Moray wrote to Lauderdale, the leader of Engagers,[30] and openly used Masonic terms, reminding Lauderdale that he intended to 'play the Mason' when he meant he was about to impart secret information. Moray would not have been likely to write thus to a non-Mason, as it would have been meaningless.

Additionally, I knew that Moray also used Masonic terms in a letter to Charles II in 1653. This suggests that Moray thought Charles II was a Freemason and would recognise a veiled reference to the Masonic character Hiram Abif. Moray did this at a time when his life depended on communicating clearly. Had Charles been made a Mason by Engager members of the Lodge of Edinburgh during his few days on the run from Gowrie House?

Perhaps Preston was right when he reported Charles II to be a Freemason. If he was, this information throws a new light on Moray's actions at the Restoration. He did not return to London with the king, but instead used his Masonic connections in Maastricht to collect up-to-date information about the intentions of the Dutch. He then went to Paris before eventually returning to London. Moray had been a senior officer in the Scots Guard and had acted as a spy for both Richelieu and Mazarin. It would have been a sensible step for him to also collect intelligence about the French intentions towards the Dutch in any future disputes.

When he did come to London in late August 1660 eyewitness Bishop Burnet reports he was greeted by Charles 'with crushing and shaking of his hand, and with as much good looks and as much kindness as I could wish'.[31] Was this a Masonic handclasp between two Brothers of the Craft?

The king certainly treated Sir Robert like a long-lost brother, giving him apartments in the Palace of Whitehall and setting up a laboratory for their shared use. Even though Moray had not been in London since 1645, fifteen years earlier, within three months he was being invited to the regular meetings of the leading Parliamentary philosophers who had recently returned to London. How did he manage to get in contact with them so easily? And how was he able to convince them of his goodwill to such an extent that they trusted him to tell the king about their meetings? Wilkins and Ward, in particular had no reason at all to trust the goodwill of a king who had just thrown them out of senior university posts.

These meetings were held at Gresham College. This institution seems to have provided a Masonic safety net for many of these philosophers. When they lost everything else, many of the founders of the Royal Society had fallen back on professorships at Gresham to offer them food, lodging and a place to meet with fellow philosophers. Moray, as a Freemason, would have known of the college Sir Thomas Gresham had established and it would have been a logical place to seek out fellow Masons – Fellows of the Craft who were interested in studying the hidden mysteries of nature and science. Moray's use of an existing Freemasonic network made sense of the whole sequence of the otherwise unlikely chain of events and gave a context for one of the early experiments generally attributed to the Society.

In 1661 Christiaan Huygens visited London and attended a Society meeting. A few years earlier he had invented a pendulum clock, as a new and accurate means of time keeping. Historian Lisa Jardine said about this visit:

> *In 1661 Huygens had visited London for the first time, and his new pendulum clock had attracted the attention of several members of the Royal Society with naval connections – particularly Sir Robert Moray and a Scottish nobleman, Alexander Bruce.*[32]

Lisa Jardine is a lively and thoughtful historian and she would not have made this comment unless she had reason to accept it as reasonable, but neither Bruce or Moray had any naval connections. I knew that Moray was a soldier and Bruce's interest in marine matters was limited to seeking continental markets for his father's coal.[33] Yet when Huygens demonstrated his new pendulum clock to Moray it was Bruce who was asked to carry out sea trials of the clock to see how accurate it was keeping time on board a moving ship. Moray was aware that the longitude of a vessel could be calculated by comparing the local time, measured from the sun, with the time at the ship's port of origin. The problem was keeping an accurate measure of the time at the originating port.

Accurate clocks were a new invention. Galileo had discovered that a pendulum could be used to measure time, but it was not until 1658 that Huygens had built the first precise (to within 15 seconds a day) pendulum clock. But pendulum clocks only keep good time when kept in a level and stationary position. Ships at sea are very rarely level and certainly not stationary. Moray had suggested that the use of a gimbal, to keep the clock level, might allow it to remain accurate enough to be useful, but the only way to test this idea was to sail about with a clock, keeping records of its accuracy. Moray needed somebody to carry out these trials. In essence he required an intrepid, and resourceful experimenter but instead he got Alexander Bruce.

Bruce was about as far from a quick-witted, adventurous, systematically scientific explorer as it is possible to imagine. Bishop Gilbert Burnet, in a kindly mood, described him as 'slow thinking'. In 1661 he was not even a fit man, he was in London, recuperating from a long-term indisposition of the stomach he had suffered from for much of his exile. His elder brother was running the family mine in Scotland and producing coal and salt which he sold to Holland and Germany. Even when he was in the best of health, Bruce's letters to Moray indicate he was a poor sailor and his inability to retain the contents of his stomach during rough sea voyages was not helped by the dyspepsia he was suffering as a symptom of his illness. Yet this slow-thinking, seasick invalid was picked by Moray to carry out sea trials with Huygens's new pendulum watches. He hardly seems the best choice to make the accurate observations and meticulous adjustments the tests demanded.

Jardine reported that the trials were not very successful. Bruce carried out a number of voyages between London and The Hague during 1661–2. It took him a number of voyages because on the first trips he was so seasick that he failed to keep the clocks properly wound and they stopped, so invalidating the test! The gimbal mounting that Huygens and Moray had devised to allow the clocks to stay upright as the ship rolled also gave Bruce trouble. His handling of one of the clocks, while attempting to wind it without removing it from its gimbal, was so clumsy that it fell from the mounting and was too damaged to continue the trial!

Far from the dedicated scientific pioneer with an interest in naval matters that Lisa Jardine inadvertently portrays, Bruce seems to be a bumbling but well-meaning amateur doing his incompetent best. If Moray was mixing with the cream of London's scientific experimenters at that time, why use the ill-fitted Bruce for such an important job?

I suggest that there are two reasons for this choice. Bruce had access to his brother's ships, which were trading salt and coal between Britain and the Continent and, along with Moray, he was a member of the Freemasonic Lodge of Edinburgh. Bruce was not a good sailor or a particularly good scientist but he was a willing and trusted supporter of Moray. Moray understood the military need to solve the navigation problem and used whatever means he could to get his idea tested.

Eventually, however, Bruce did produce enough evidence to show that a suitable clock could be used to measure longitude with a useful degree of accuracy. Fortunately Moray was able to use this evidence to convince the king that a proper naval trial should be carried out and a year later the far more competent Captain Robert Holmes was given the job of testing the clocks on longer voyages.

Although Huygens's pendulum watch was an improvement on dead reckoning, it did not prove to be good enough to solve the longitude problem. The improvement of marine watches was a matter that would be of ongoing interest to the early Royal Society, but it would be left to Harrison to finally perfect the seagoing chronometer, one hundred years later.

Conclusion

A more detailed study of the events surrounding the Restoration had convinced me that Charles II had been a clever politician who was well able to charm people into helping him. He also had significant problems with his navy, a major threat to his overseas colonies and no money to help solve the technical problems of improving his fleet.

Sir Robert Moray's actions made sense in the light of the Masonic and military connections he had with a number of important players in the politics of the day.

The strange events surrounding the founding of the Royal

Society were starting to make more sense in the context of a Masonic connection between some its founders. Perhaps I could now start to make sense of the complicated saga that was Moray's quest for a Royal Charter.

CHAPTER 9

The Royal Charters

Sir Robert Moray had been considering how the group of philosophers might best be converted into an organised institution with the necessary safeguards; he knew that bitter opposition was to be expected from both ecclesiastical and political bodies to any development of natural philosophy in this country. The only effective safeguard which it was within the philosophers' power to employ would be to obtain from the king their incorporation by means of a Royal Charter.[1] Sir Henry Lyons, FRS

WHY SIR HENRY ASSUMES that a group of Parliamentary misfits and outcasts should have the power to command a Royal Charter from the king, in order to protect them from the criticism of the establishment, is something he never explains. He is, however, right in his analysis of the effect a Royal Charter would have on the fortunes of the fledgling society; but my close examination of the political power of those first twelve members at the 28 November meeting had shown that the majority had no political influence at all.

Of these twelve men only Sir Paul Neile or Sir Robert Moray would have been in any position to arrange an immediate audience with the king. Of the two Moray was best placed to know Charles well enough to speak to him about a possible Royal Charter. Neile had been a courtier to

Charles I and had been reinstated as a Gentleman of the Privy Chamber some four months earlier. Charles had re-appointed him to the same post he had held under the previous king and after that Neile did little else of note, apart from feathering his own nest by investing in the Hudson Bay Company. He seems to have been rather a nonentity. During the Civil War he hid away and did little scientific work apart from grinding lenses in his spare time and he had no contact with the exiled king. His main contribution to the Royal Society seems to have been financial.

Moray was the mover and shaker who had actively served the king during his exile. He returned to a vigorous Royal welcome and had been given his own private house in the grounds of the palace of Whitehall. Despite Sir Henry Lyons's comments in the opening quote of this chapter, there was only one person able to deliver a Royal Charter to this ramshackle bunch of philosophers and that man was Sir Robert Moray.

But I was still not clear as to how and why he did this and so I determined to look more closely at the political background and the timing of the events surrounding the first meetings of the Royal Society.

On the Wednesday of that first meeting (28 November 1660) the king was not in London. He had set off to Dover to meet Henrietta Maria, who was due to arrive back from exile in France. Charles was not in a particularly conciliatory mood towards ex-Roundheads. Two weeks earlier he had stood by the gallows in Charring-Crosse and applauded while four of the men who had condemned his father to death met their own ends (quite literally) by being hanged, drawn, quartered and then their parts jumbled into wicker baskets to be paraded round the city.[2] His feelings towards the supporters of Cromwell would not have been soothed by the councils of his mother, who never forgave the English for murdering her husband. Charles could hardly have avoided discussing the executions during the journey back to London, which took until Monday 3 December 1660.

Early in the afternoon of Wednesday 5 December 1660 Sir Robert reported to the second meeting of the Royal Society: 'the King had been acquainted with the designe of the meeting. And he did well approve of it and would be ready to give encouragement to it.'[3] The implication of this statement is that the king had nothing else to do and so Sir Robert sat down and discussed his plans for totally reorganising the scientific methods of the country and the king was so impressed he insisted that he must be part of this great experiment and support it. A closer look at Charles's actions between his return late on Monday and Sir Robert's announcement on Wednesday afternoon suggests a different scenario.

Sir Robert only had a small window of opportunity in which to speak with the king. The king had been escorting his demanding and overbearing mother on her journey from Dover to London and had not returned to Whitehall until late on Monday. On that same evening Pepys's *Diary* reports that the king had been consulted about the problems that the Admiralty had in finding enough money to pay the wages of the seamen, saying, 'the king doth take very ill (the proposition that paying the seamen's wages should be postponed)'. This betrayal of good faith to the men of the fleet was such a worry to the king that he dispatched his brother to the Admiralty the following morning to see what could be done.[4] On the Tuesday morning the king attended the House of Lords to hear Parliament vote favourably on the proposition that the 'bodies of Oliver, Ireton, Bradshaw and Pride should be taken out of their graves in the Abbey and drawn to the gallows and there hanged and buried under it'. This vindictive action towards Cromwell in particular upset Pepys who said of it, 'It doth trouble me that a man of so great courage as he was should have that dishonour, though otherwise he might deserve it enough.'[5] What is very likely is that on his return Henrietta Maria would have insisted on an immediate report on the method he intended to use to take his revenge on the regicides.

Despite all these distractions, less than twenty-four hours later, Moray was confidently quoting the king's approval and support for a project which had been overseen, quite openly, by Oliver Cromwell's brother-in-law. The king seemed to be playing a very two-faced role: rejoicing in revenge while cynically using out-of-work Roundheads to work on improving his run-down and under-funded navy.

The first resolution passed by the new committee was about the frequency of the meetings and the payment to be made:

> *Wee whose names are underwritten, doe consent and agree that wee will meet together weekely (if not hindered by necessary occasions), to consult and debate concerning the promoting of experimental learning. And that each of us will allowe one shilling weekely, towards the defraying of occasional charges.*[6]

At this second meeting a committee was set up to discuss the content of a Constitution for the Society and to consider where the regular meetings should be held. It had been suggested that the group should move from Gresham College to the College of Physicians. By 19 December the committee, which included Brouncker, Moray, Neile, Goddard, Matthew and Christopher Wren had resolved that:

> *the next meeting should be at Gresham College, and so from weeke to weeke till further order.*[7]

To understand what Moray was doing it is easier to consider the two different, but complementary, strands of action he was following. One line was administrative and considered what the Society was to be called, who was to become a member and how it was to be run and financed. The other strand concerned the scientific work which was to be tackled.

The meeting on 19 December 1661 confirmed the importance of John Wilkins in organising the scientific effort of the new group. *The Journal Book* records:

> *That Dr Wilkins and as many of the Professors of Gresham College, as are of this Society, or any three of them, to be a Committee for the receiving of all such experiements. (fit for the advancement of the generall designe of the company.)*[8]

Of the seven Gresham professors, at that time, five were members of the new Society. It is also clear from the minutes that Wilkins considered Moray to be linked with the Gresham professors as his minute of 16 January 1661 says:

> *Those of this Society who belong to Gresham College, together with Sir Robert Moray, and as many others of the Company as will meete, will be a Committee about Magnetical enquiries.*[9]

Here for the first time Moray, a Freemason of twenty years standing, is specifically linked with Gresham College, which is reputed to have been a Masonic foundation.

By 20 March 1661 Gresham was providing not only a public meeting room but another 'repository to keep Instruments, Books, Rarities, Papers and whatever else belonges to them'.[10] The Professors of Gresham had also been given the job of preparing the meeting room for experiments. *The Journal Book* for 20 March 1661 says:

> *That the gentlemen of Gresham College be Overseers for accommodating of the roome, for the Society's Meeting.*[11]

Interestingly this is yet another instance of the Society using

Masonic terms in addition to that of the Fellow. (Overseer is a Masonic rank within the Mark degree.)

Margery Purver commented on how well the Society fitted into the ethos of Gresham College:

> *It is perhaps not irrelevant to emphasize that its character [the new society], which had been determined at Oxford, experienced no change on its incidental contacts with Gresham College.*[12]

Knowing what I now knew about the Masonic origins of both groups this insight seemed much less surprising to me than it had done to Purver. The Society continued to meet and debate experiments at Gresham College until they were driven out by the Great Fire of London in 1666.

But, to return to the story of the Royal Charter and my attempt to unravel Moray's political plan for his group of philosophers. On 12 December 1660 the new society drew up a set of Rules of Conduct. These are very similar in form to the Antient Charges which Moray would have been familiar with from his contacts with Scottish Freemasonry and he may well have modelled the Society's first rules on what he knew of the Charters and Charges of Freemasonry. Moray would have known from experience that a full set of rules made for easier management of lively and strong-minded individuals. He may well have remembered the part of the ritual which urges a Master of a lodge to use the rules to ensure good government:

> *In all cases of trouble and difficulty you are reminded to consult our Antient Constitutions for there is scarce a case of difficulty which they will not offer you guidance upon.*

The rules, which the new Society set down in writing, covered: the election of Fellows, the election of Officers and servants of the Society, the keeping of minutes and the payment of fees. In another similarity to Freemasonry the Society decided to have three officers to rule it, an idea Moray would have been familiar with from the concept of a Master and two Wardens who govern a Freemason's Lodge. Even the method of voting, using a positive and a negative token with a box for positive and different receptacle for negative tokens is an exact copy of the method of secret ballot which is still used to this day in Freemason's lodges. Moray seems to have drawn freely on his knowledge of the rules that his Mother Lodge of Edinburgh had developed to organise its affairs and he adapted them to place the new Society on a firm foundation.

Now Wednesday meetings at Gresham became the norm. The rules provided for the monthly election of a President to preside over the meetings and on 6 March 1661, the rules having been passed and accepted by the members, Sir Robert Moray was elected as the first President. He was to take this role a further eight times until enactment of the Royal Charter which made Lord Brouncker the first long-term President.

Preparing the First Charter

The Journal Book of the Society shows that Moray was the driving force behind building an organisation which was structured well enough to be able to request a Royal Charter. Sir Henry Lyons says of this period:

> *Moray was extremely active throughout the year 1661 and this was recognised by his colleagues who elected him to the presidency of the new Society oftener than any other of its members. He undoubtedly had much to do with the preliminary drafting of the First Charter.*[13]

On 21 June 1661 Moray wrote to the scientist Huygens, saying that he was working hard to set up the new Society on a good sound basis. Moray was by this time using all his political skills to make sure that the king was convinced about the aims and ambitions of the society. He had been paying particular attention to John Evelyn, who was another confidant of the king. He had taken Evelyn to see Robert Boyle's experiments with air pumps and diving bells on 9 March 1661.

Boyle was continuing to carry out experiments with both compressed air and with vacuums. He had just discovered, and demonstrated to the new society, that an enclosed volume of air will act just like a spring. If you squeeze air into a small space it will try to spring back to its original size and push outwards as it tries to do so. By measuring the force on a piston and the distance it would move Boyle showed that if the volume of the gas was halved, its pressure would double. This inverse relationship is still known as Boyle's Law. With this experiment Boyle laid down the basis of a discipline we now call thermodynamics. Boyle's Law leads to a relationship between volume, pressure and temperature, known as the General Gas Law and from these ideas have developed the whole theory of heat engines, of which the most dramatically successful example is the internal combustion engine which powers our motor cars.

It is easy to imagine Robert Boyle as a systematic and alert scientist, carefully studying and writing down the details of a series of experiments before announcing his discovery of a new law of nature but the reality was a little different. Boyle had a temperamental and unreliable air pump and no clear idea of exactly what he was trying to find. As this description from John Evelyn's diary from 7 May shows, Boyle often carried out dramatic demonstrations just to entertain visitors, especially Royal visitors:

> *I waited on Prince Rupert to our Assembly, where were tried several experiments in Mr Boyle's vacuum: a man thrusting in his arme, upon exhaustion of the ayre had his flesh immediately*

> *swelled, so as the blood was neere breaking the vaines and unsufferable; he drawing it out, we found it all speckled.*

Moray seemed to be encouraging two things, development work on matters of naval importance and a series of showy displays of scientific curiosities to encourage the support of the nobility. Not all such experiments were of a type we would recognise as scientific today. For example when the Duke of Buckingham was admitted a Fellow on 5 June 1661 he presented the Society with some horns said to be those of a unicorn. In 1661 it was a well-accepted fact that a circle of unicorn's horn could act as an invisible cage for any spider. The learned scientists of the new society took the unicorn's horns, provided by Buckingham, and watched while Robert Hooke placed a spider within their circle. It walked away without as much as a second glance at the magical container. To be fair, the Duke had not actually provided the society with real unicorn's horns. He had inadvertently given them horns taken from the white rhinoceros. The test, however, has never been repeated with real unicorn's horn. This may in part be because of the mythical nature of the unicorn and the subsequent difficulty in procuring the genuine item!

These side-show experiments of Robert Boyle's were part and parcel of the early meetings. Demonstrations of fast-acting poisons on small animals were always entertaining for the noble Fellows, but now and again observations of strange trials actually founded real science. If Boyle had not noticed the swelling of the man's arm he would not have started to think about air pressure and its effects. And without a study of the behaviour of gases within cylinders we would have no engines to drive our cars.

But Moray wanted to keep the king's mind focused on the Society, no easy matter when the king's thoughts, and often actions, were so easily diverted by the nearest available lady.

When Evelyn wrote his *A Panegyric to Charles the II*, which he published to celebrate Charles's coronation on 23 April, he said within it:

> *Nor must I here forget the honour You have done our Society at Gresham College . . . For You is reserved being the Founder.*

Moray was making sure that he used every means, including flattery, to keep his Society in the forefront of the king's mind.

That Moray based his thoughts on how the Society should be ruled on his experience of Freemasonry is clear from the rules and attitudes that were adopted. By September 1661 he was ready to send a draft of the proposed constitution to the king for consideration. He was also making sure that the king's research needs were being met by the group. During this period the Society looked into the accuracy of compass needles, the use of Jupiter's moons to determine longitude (Charles had already funded a telescope for Moray's use at Whitehall and on 3 May 1661 Evelyn's diary records the king took part in some observations of Jupiter) and commissioning that famous model, built by Christopher Wren, to test out the idea of Laurence Rooke to use the moon as a universal sundial. The fledgling society was focusing its main efforts on the needs of the Navy but in the process was developing a new approach to science. Modern science tries to follow a systematic procedure for developing its ideas. There are three important steps in the process: *observation*, *prediction* and *control*.

First *observe* the phenomenon you are trying to explain, name its parts, describe its functions and make sure that you can record exactly what happens. Sometimes this step will involve developing new measuring instruments, such as telescopes to view the stars, or microscopes to observe the detail of cells. On other occasions it will involve making a more accurate clock to count fractions of a second, or a more accurate calendar to time the years and the seasons. Frequently it will involve carrying out experiments to see what will happen, such as pumping away the air from around a man's arm to see if it swells up, or stuffing snow into a dead chicken to see if it rots more slowly. Perhaps it involves collecting and classifying plants into different types, which can then be named, or seeing if trees grow faster when scattered with saltpetre. The founders of the Royal Society did all these things, and more, at a time when most 'scholars' never bothered to observe anything.

Descartes, the philosopher, who was a contemporary of many of these founders, had based his system of mechanics on two principles: the identification of space with matter and a view of motion which only allowed for moving bodies to move from contact with one body in order to make contact with another. His view of space has survived with his system of Cartesian coordinates which is a method where any point in space can be named by counting along two axes at right angles to describe any position (perhaps the best known example being the way in which the position of a piece on a chessboard is described by counting the rows and columns). But his ideas on motion came right out of his head. He said that a body only moved in order to make contact with another body, so the opposite to motion is rest and all motions results in impacts between bodies. Try telling this to the snooker player who has just missed potting the final black ball by playing a non-contacting foul stroke! But, like Aristotle before him, Descartes never let observation get in the way of a good theory.

As the new observational attitude to science that the Society promoted spread throughout the academic community, scientists took advantage of the new-found freedom to look at things which religion had previously held sacred. Fellow Number 90, Thomas Willis, who joined on 13 November 1661, explored the anatomy of the brains of his patients who died, and founded the modern science of neurology. He was ejected from the Society for failing to pay his subscriptions and then, two years later, had to rejoin as F153. When he moved from Oxford to London he found he needed the support of the Society to continue his experiments. From then on he paid up more regularly!

The fundamental change in attitude towards observation, which the Royal Society caused, can be seen when in 1668, John Wallis, Christopher Wren and Christiaan Huygens were asked by the Royal Society to devise a mathematical method for working out the motions of projectiles. Wallis, perhaps the most celebrated of all theoretical thinkers of the time, and the father of modern algebra, wrote when accepting the commission, 'Experiment will be the best judge of our deliberations.' This was put into practice when Charles asked them to investigate why, when a frog was put into a brimming glass of water the glass did not overflow. The point was quickly proved by taking a frog and a glass of water and noting that the glass did overflow!

After observation, the next stage of the process of science is that of *prediction*. Here it is necessary to build a theoretical model to explain the observations and then use the model to predict the outcome of an, as yet untried, experiment. If the prediction is accurate the model can be considered to be a reasonable representation, but if the model fails to predict the outcome then it must be discarded or rebuilt until it can predict. It is a harsh reality of this scientific method that no matter how many times a model has worked in the past, if it fails once it is wrong and should be discarded or changed. I must add here that this is a counsel of perfection and does not always happen. Scientists, being only human, hate discarding old, comfortable and familiar theories.

The scientists of the Royal Society put forward predictions and tested them against reality. They built diving bells, twin-hulled ships, wind-driven carriages, spring-driven clocks and new sorts of pistols to test the predictions of their new theories. Perhaps the most famous example of the application of this principle of observation followed by prediction is the story of Halley's Comet. Prior to the Royal Society comets were not understood. By fostering the development of accurate telescopes, precise clocks and systematic recording of observations, the Fellows of the society built up a database of sightings of comets. From the detail of the directions of the sighting they created theories of how comets moved until finally, Edmund Halley, using the observations of John Flamsteed, and the theories of Isaac Newton, predicted the regular return of the comet which bears his name.

The final stage in the process of science is the step of *control*. Once you can describe a process, and can predict its outcomes, you can then hope to intervene and adjust the outcomes to be as you want them to be. This is the step of control. Modern science has given us such a high degree of mastery over our world that we have almost forgotten what the pre-scientific world was like. When Robert Hooke first started to draw the minute creatures he saw through his microscope he was astonished at what he saw. Hooke amused the noble members of the society by supplying them with drawings of the tiny creatures he observed through his eyepiece and so helped keep the funds flowing. However, looking closely at plants and at human skin, while he was making these amusing drawings, he saw tiny subdivisions in the basic structure. He described these as looking like the small cells of monks, grouping together to form a cloister. And so the name cell was born for a basic unit of biological life. The first step on the road towards the cloning of living beings had been taken. The name that Hooke used survives in the same context today. The study of biological cells stretches directly back to Robert Hooke and his first microscopic studies of tissue. Hooke's observation of the basic biological unit and his decision to call it a cell still echoes through the literature of science surviving in Hooke's picture of the spaces he saw in living tissue as 'cells' (meaning small empty rooms).

The name stuck and Hooke went on to observe living cells; later scientists predicted how these cells could be manipulated and, eventually, Wilmut and Campbell achieved control of the technology of 'cells' and created Dolly, the first cloned sheep[14]. Here is yet another development in modern science that can be traced back to the founders of the Royal Society. But all this started with Moray's plan to found a society to help in sorting out an ailing navy. Petty, Goddard, Digby and Christopher Wren were formed into a committee to consider and improve shipbuilding. Brouncker was encouraged to discuss his ideas on improving the recoil action of naval guns. Boyle and Brouncker worked together on producing a set of questions about the changing nature of the seas which were to be answered by records kept during a naval expedition to Tenerife.

As a direct result of this the Society published a paper, in 1666, called *Directions for Seamen going into the East & West Indies*. As historian C A Ronan pointed out:

> *The idea of these 'Directions' was to ensure that the Society had access to various kinds of information which mariners visiting the East and West Indies could obtain and which could be of real use for furthering geographical, navigational and other scientific studies. Copies of the journals would, it was proposed, be deposited with Trinity House as well as with the Admiralty, and the Society should have a right of access to them.*[15]

This simple collection of information was a tremendously important step in developing the scientific method. Today we take for granted the vast amounts of data that are automatically collected, stored and published. But this tradition was started by the early scientists of the Royal Society.

Evelyn was encouraged to develop his techniques for etching, illustration and map-making and he also cooperated with

Digby to look at the problems of horticulture and the cultivation of trees.[16] The reason the Society sponsored the construction of an experimental diving bell was to seek a practical naval use for Boyle's new air pump.

The instigator of this diving bell experiment was Sir Jonas Moore. When Charles II married Catherine of Braganza he received Tangiers as part of her dowry. Moore was put in charge of improving the fortifications of the port and was aware that he would need to carry out a great deal of underwater construction work. The only way of doing this was using divers who held their breath as they worked. The time they could work was extremely limited and Moore developed something, which was described to the Royal Society as 'an engine for staying two or three hours under the water in'. In fact it was a diving bell, simply a heavy cast bell, such as might be hung in a church tower. It was lowered on to supporting stones on the sea bed. The bell held a bubble of air. The divers could work for a while before slipping under the bell to snatch a few breaths of air before continuing. This device had the disadvantage that it had to be regularly lifted out of the water to freshen the air it contained. Sir Jonas contacted the Royal Society to see if they could adapt the air pump, which Robert Boyle had invented, to keep the air fresh.

The air pump, which only Robert Hooke could operate reliably, had previously only been used to provide interesting and entertaining experiments for the noble members. Small animals would be placed inside a glass jar and the air removed to see what happened. A wasp, which found itself unable to fly, got extremely annoyed and caused great consternation when released to see if it regained its ability to hover. It did!

If the pump worked well, most animals quickly died, if it sprang a leak they simply choked a little. On one notable occasion Hooke built a large glass fronted airtight container and stood in it while an assistant pumped out its air. Fortunately the pump was not having a good day and Hooke survived. He reported that as the air was pumped out he felt pains in his ears and chest. We are not surprised at this result, but in 1661 this was pioneering knowledge about the effects of oxygen on life.

However, to return to Sir Jonas's request for assistance with the problem of naval divers, an airtight lead helmet was made and connected by a tube to the output side of a higher capacity air pump. Hooke recalled in his diary how Boyle dispatched him to London to collect 'a barrel and other parts of the engine which could not be made in Oxford'. In May 1664 Sir Robert Moray reported to the king that the trial had been unsuccessful in so far as the diver had only been able to stay under water for four minutes. He added the disturbing comment that the man had been 'ordered to continue practising'.[17] Possibly in the hope that enough practice would enable the hapless diver to avoid breathing at all! This work, however, was not entirely without its benefits as Robert Boyle published a book about his reflections on diving called *Hydrostatical Paradoxes* in 1666 in which he laid the foundations of the modern science of hydraulics.

Moray was using his political skills to demonstrate to the king that supporting the new Society would result in useful inventions to improve the strength of the navy. But he also provided small scientific delights to amuse the king. When the king heard of the strange and amusing creatures to be seen under Hooke's microscope he showed a keen interest. Moray instructed both Christopher Wren and Robert Hooke to produce a series of detailed micrographic drawings of small creatures for the king's delight. To this end they drew a louse, many different types of flea and sections of a fly's wing. The king loved the drawings and wanted more. Eventually Hooke published the collected microscopic drawings in a book. It was entitled *Micrographia*, and he wrote it in English (rather than Latin) and illustrated it himself. The book contains some of the most beautiful drawings of insects and the minute structure of feathers and fish scales ever seen.

It was, however, after the first draft submission of a charter, that Moray saw an opportunity to use the Society to help Charles in a more important area, that of foreign policy. In November 1661 John Winthrop, son of one of the Founders of Massachusetts, and now Governor of Connecticut, had come to London looking for a Royal Charter for his colony. The chain of command from England slowed decision-making in Connecticut and stifled initiative. The degree of independence, and local administrative autonomy a charter would afford was a logical step forward, especially now the tight grip of Cromwell's Protectorate had passed. Moray, who had met Winthrop at Oxford in 1643, invited him into the Society and once more took on his old role of negotiator, helping Winthrop get the charter he sought. Winthrop was also appointed 'the Chief Correspondent of the Royal Society in the West'. Moray was well aware of the ambitions of the Dutch and he seems to have made sure that he would have a regular means of keeping up to date with the unfolding threats of Dutch expansionism in North America.

By now the inner circle of scientists were openly calling their creation 'The Royal Society'. In its actions and intentions it was harnessing a study of nature to support the needs of the king in general and his weak navy in particular. The first public record of this title was, once again, made by John Evelyn in November 1661 when he used it in a dedication to Edward Hyde, Earl of Clarendon and Lord Chancellor. It is tempting to speculate that Moray suggested to Evelyn that it would be a good idea to butter up Hyde to prepare the ground for a successful Royal Charter to make the title a reality. After all Hyde would have been quite able to quash the idea had he not been flattered into accepting it as a *fait accompli*.

Robert Boyle also commented on the sensitivity of the choice of name for the Society saying, 'the Illustrious Company that meets at Gresham College have hitherto suspended the Declaration of themselves as a Society'.[18] It seems that there was some dispute about what they should be called. 'The Royal Society' with its overtones of total support for the king was not a unanimous choice.

Christopher Wren wrote a long preamble to the draft charter which included the following statement:

> *And whereas we are well informed that a competent number of persons, of eminent learning, ingenuity, and honour . . . have for some time accustomed themselves to meet weekly, and orderly, to confer about the hidden causes of things, with a design to establish certain and correct uncertain theories in philosophy and by their labours in the disquisition of nature, to prove themselves real benefactors to mankind; and that they have already made a considerable progress by divers useful and remarkable discoveries, inventions and experiments in the improvement of mathematics, mechanics, astronomy, navigation, physic, and chemistry we have determined to grant our Royal favour, patronage, and all due encouragement to this illustrious assembly, and so beneficial and laudable an enterprize.*

This preamble was not eventually included in the final draft of the charter but Sir Henry Lyons says that it is 'valuable as a description of the Society's aims as understood and described by one of the most able of the small group of men who founded it at Gresham College'.[19] And it shows a remarkable similarity to the aims described in the rituals of the Fellowcraft degree of Freemasonry.

At a meeting held on 18 September 1662 the members present agreed on the draft for their Royal Charter and Sir Robert Moray was asked to present the petition to the king. At this time it would appear that the title of 'The Royal Society' was chosen to appeal to the king. Be that as it may, a favourable reply was returned to the fellows on 16 October 1662 when *The Journal Book* records that Sir Robert reported to the meeting that:

> *hee and Sr Paul Neile kiss'd the King's hands in the Company's Name.*

The meeting went on to ask Sir Robert to:

> *return most humble thancks to His Majesty for the Reference he was pleased to grant of their Petition.*

Sir Robert then reported the king's intention to join the Society saying:

> *and to this favour and honour hee was pleased to offer of him selfe to bee enter'd one of the Society.*

Matters of State were moving on very rapidly for Charles; he was coming under greater pressure to go to war with the Dutch. His sister Mary, who was the widow of the Prince of Orange, had been keen to press her claims for guardianship for their son, the young William of Orange in order make him the next Stadholder. When Charles had been restored he had been prepared to tell Johann De Witt, the Dutch Minster of State, that he was more interested in Dutch support for English policies than in his nephew's claims to be Stadholder of the Netherlands.[20]

Then Mary of Orange died while visiting her brother in England, and her death upset the status quo. She left guardianship of young William of Orange jointly to the Queen Mother, Henrietta Maria and Charles himself. She made no mention of either the State of Holland or her mother-in-law, Amelia, dowager Princess of Orange. Needless to say the Dutch refused to confirm the guardianship of Charles, and Amelia joined forces with De Witt. This snub annoyed Charles and he was now far more open to persuasion that it would be in the commercial interests of England for the Dutch to be defeated in a European war.[21] He was encouraged in the attitude by his surviving sister Henrietta Anne, sister-in-law of the king of France. She wrote to him:

> *It is with impatience that I can endure to see you defied by a handful of wretches and it is perhaps pushing glory a little too far, but I cannot help it, and everyone has his own humour and mine is to be very keenly alive to all your interests.*[22]

But if he was to have any chance of winning a war with the Dutch, Charles needed to make progress in all matters concerning shipbuilding and navigation. Maritime historian Ralph Davies said of this period:

> *It was happy for him that the English shipbuilding industry made prodigious advances during his reign. At first the industry was hard put to meet the rise in demand for vessels following the settled times of the Restoration, and there had to be large-scale buying of ships from abroad. However, two acts of Parliament put an end not only to foreign-owned ships carrying English trade but also to the employment of foreign-built ships. The result was a speedy advance in native British shipbuilding.*[23]

But John Evelyn reported that in August 1662 he had been rowing down his local creek when he came upon two warships lying without a soul on board and everything stolen.[24] This was not really so surprising since almost every wage payment connected with the navy and its administration was several years in arrears. Charles was forced to sell Dunkirk to the French for five million Livres to help relieve his burden of debt.

With such pressure on him to improve both the Merchant and Military fleets Charles rapidly grasped at the technical aid Sir Robert was offering him through the new Royal Society. Charters, after all, cost less than ships and the new fellows had no arrears of pay owing to them!

The privileges Charles offered to the Royal Society in the First Charter of 1662 were of two types. Firstly there were internal privileges, which just related to the way the Society was guaranteed the right to manage its own affairs without interference from the State. These included the right to appoint a President, a Council and Fellows, who would decide the way in which the Society was to be run. The only restraint the charter placed on the actions of these officials was that their behaviour had to be 'reasonable and not contrary to the laws of the realm'. But within this restriction they could freely make whatsoever laws, statutes and orders they wished.

The external privileges of the Society were concerned with its corporate identity and the powers that became vested in that Society as a company. The most important privilege was that of continuity. Once created the Royal Society became immortal. Its President, Officers and Fellows could live and die but the Society itself continued to exist. It had exactly the same rights and obligations in law as any other subject of the king. In addition it had the right to meet either within London or within ten miles of the city but perhaps one of the greatest privileges the First Charter gave it was the right to print and publish and to carry out correspondence with foreigners, provided it was on scientific matters. These were important rights in the seventeenth century when censorship by either the Church or Crown was taken for granted.

One of the more useful, if somewhat macabre, rights the

Society was given was the right to demand possession of the bodies of executed criminals, which its members could then dissect as they wished. It also had the right to build colleges either in London or within ten miles of it. To ensure ongoing harmony within this new corporate body the charter made provision for persons of high position to be appointed as arbiters if the Society should not be able to resolve its own disputes to the satisfaction of its fellows.

The name that the First Charter conferred on the organisation was a simple one: 'The Royal Society'. This name, which had been agreed by Sir Robert during his negotiations does perhaps sum up what he saw as the purpose of the Society. It was to support the king in the provision of technical advice, and from the early subjects the Society tackled, there was a particular interest in the Royal Navy and its problems.

The Second Charter

The First Charter was given the Royal Seal on 15 July 1662, some eighteen months after the initial meeting of the Gresham Twelve. As during this period the king had also set up a new government, celebrated a coronation and also married, this was remarkably swift progress. But as the saying goes, 'more haste, less speed'. Had Sir Robert really merged this mixed bunch of Parliamentarian technologists and Royalist fundraisers into a society which could admit its true purpose was to support the Monarchy? It would appear that he hadn't, because when the charter was read out to the Society on 15 Aug 1662 the Fellows kindly thanked the king but then asked him to consider granting a further charter.

Margery Purver points out that:

> *Despite the formal expressions of pleasure and gratitude, the Society was not satisfied, and within a few months announced that alterations were to be made to the Letters Patent. It has*

often been stated that the Royal Society obtained 'further privileges' in the Royal Charter of 1663, without specifying what these were. But the only privilege which the Society did not already possess was the grant of a coat-of-arms, which might well have been made separately, without the considerable labour of preparing and passing another comprehensive charter, particularly so soon after the original Charter.[25]

Two important things were changed, however: these were the name of the Society and its relationship with the king. The name assigned to the Society under the First Charter was The Royal Society and no mention was made of the king as Founder. The Second Charter, which Sir Robert pushed rapidly past the Royal Seal on 23 April 1663 and presented to the Royal Society on 13 May 1663, was awarded the title 'The Royal Society of London for Promoting Natural Knowledge'; also, within the new Letters Patent the king made quite clear his relationship to the Society saying, 'of which same Society we by these presents declare Ourself Founder and Patron'.[26] This new form of words must have finally satisfied the Parliamentarian 'techies', who were the scientific driving force of the group, because after this the Society developed rapidly. The king was so pleased to be acknowledged Founder, without having to provide any funds, that he gave the Society a Mace. On 23 May 1663 Charles sent a warrant to the Royal Jewel Office to make for the Society 'one gilt Mace'.[27] In the Second Charter Charles also makes clear that he links the formation of the Society with his vision of an England which is strong abroad, and by implication has a powerful navy to enforce this policy. His words say:

We have long and fully resolved with Ourself to extend not only the boundaries of the Empire, but also the very arts and sciences. Therefore we look with favour upon all forms of

> *learning, but with particular grace we encourage philosophical studies, especially those which by actual experiments attempt either to shape out a new philosophy or to perfect the old. In order, therefore, that such studies, which have not hitherto been sufficiently brilliant in any part of the world, may shine conspicuously amongst our people.*[28]

On 13 August 1663 the Society formally recorded that:

> *Sir Robert Moray should be thanked for his concern and care in promoting the constitution of the Society into a Corporation.*[29]

Sir Robert evidently thought that by now he had finally got his disparate factions working together because immediately afterwards he wrote to Christiaan Huygens saying that he would now be able to work more effectively for the establishment of the Society than had hitherto been possible and that the constitution of the Society was now as the Fellows intended.[30]

But the Fellows had changed very little. They had altered the name to reflect that their interest was wider than simply supporting the king, but they had also formally acknowledged the king as their Founder. (It does seem a little ironic that today they are still widely known by the name they rejected, The Royal Society.) But they had also changed some council members. The First Charter had named the officers who were to be appointed to set up the Society. These were:

> *Viscount Brouncker, President. Members of Council, Sir Robert Moray, Robert Boyle, William Brereton, Sir Kenelm Digby, Sir Paul Neile, Henry Slingby, Sir William Petty, John Wallis, Timothy Clarke, John Wilkins, George Ent, William Erskine, Jonathan Goddard, Christopher Wren, William Ball, Matthew Wren, John Evelyn, Thomas Henshaw, Dudley Palmer and Henry Oldenburg.*

There are nine of the original Gresham Twelve in this first Council. The missing three are Alexander Bruce, Laurence Rooke and Abraham Hill. Bruce had by this time inherited his elder brother's estate in Scotland and had returned to Culross to manage it. Rooke had recently died, so of the original Gresham Twelve only Hill had been left out. Of the other twelve members ten were taken from the list drawn up at the first meeting. Two outsiders had been brought in. Dudley Palmer was a lawyer but William Erskine was a senior member of the Court of Charles II and Cup-bearer to the king.

However, in the Second Charter, when the fellows had the council 'they intended' one of the Gresham Twelve had been rejected; i.e. Christopher Wren, in favour of another member of the Gresham Twelve, Abraham Hill. And the second list choice John Wallis had been dropped in favour of Sir Gilbert Talbot, Master of the Royal Jewel Office to King Charles II.[31] The importance of this change of membership is easier to understand once it is realised that the charter gave total control of the Society to the Council without any need to consult the Fellows. Sir Henry Lyons says of this:

> *The Council could at any time revise, or revoke, any existing statute, or adopt a new one under the powers conferred on it by the Charters; no reference to the general body of Fellows was necessary except when the admission of new Fellows, or the appointment of members of Council was concerned.*[32]

Now by this time Christopher Wren had been claimed as a Freemason, by William Preston. Meanwhile John Wallis was a known close associate of Freemason William Hammond, a London lodge brother of Royalist Freemason Elias Ashmole. The other member of the Gresham Twelve who had also been excluded from the Council was Alexander Bruce, also a Freemason and a Royalist. Did the Parliamentarians feel that Sir

Robert Moray would exert too powerful a grip on the 'Royal Society' if too many of his brother Royalist Masons were on the Council?

There is an additional piece of evidence supporting this hypothesis and that concerns the Coat of Arms of the Society. The only other major change between the First and Second Charter is that the Second Charter granted the Society a Coat of Arms. This had been the subject of some debate. Eventually the king had chosen the shield and the Fellows the crest, supporters and motto.

The final form is described in the Second Charter thus:

> *In testimony of Our Royal favour towards them, and of our peculiar esteem for them, to the present and future ages, these following blazons of honour, that is to say: in the Dexter corner of a silver shield our three Lions of England, and for Crest a helm adorned with a crown studded with florets, surmounted by an eagle of proper colour holding in one foot a shield charged with our lions; Supporters, two white hounds gorged with crowns; to be borne, exhibited, and possessed for ever by the aforesaid President, Council and Fellows, and their successors, as occasion shall serve.*[33]

But Ashmole records that he had also proposed a coat of arms which had been rejected. It used a shield which was white on the upper half and black on the lower (this is known to Freemasons as the Beauseant and is used to decorate the lodge in certain degrees of Freemasonry). In the Dexter corner of the Masonic shield Ashmole has placed the Royal Arms of the House of Stuart. In the forefront of the shield he has placed a hand holding a plumb-rule. (This is the symbol of a Warden of the Craft, to this day.) This extremely Masonic Coat of Arms was rejected by the king in favour of the more neutral one. Was this the beginning of a move away from the Freemasonic beginnings of the Society into something new?

The whole matter of the rejection of the First Charter and the changes in the Second seems to have revolved around the links with the king and the make up of the council. Margery Purvey commented that the changes seemed unimportant:

> *Except for the addition of the coat-of-arms there is very little difference in the practical terms of the two charters. Such differences concern only minor matters and relate to internal administration, which the Society already had the right to alter at will.*[34]

However, if Moray had reported to the king that there was discontent among the active scientists with the degree of control being given to long-standing Royalist supporters, this could easily explain why these changes were made. It would also explain why Moray, who had taken the President's role more than any other member in the interim period, stood aside for the relatively unknown, but neutral Brouncker to take the Chair.

Conclusion

The support of the king had to be pre-planned as he had virtually no time to be consulted. As only Sir Robert Moray was involved in the negotiations at this stage he had to be the man behind the Society. He had used his Masonic background to make contact with deposed Parliamentarians who had the skill to attack the technical problems of the king's navy.

His diplomacy had worked quite well up to the moment the First Charter was read out to the Fellows and they realised Moray was linking them more closely than they wanted to the king. A compromise was reached resulting in the Second Charter, which reduced the influence of Moray's supporters. The Royal Society began to take on an independent life of its own.

I was beginning to suspect that I had found the period when a Freemasonic organisation began to mutate into something I could recognise as a scientific society. To check out this idea I decided the next stage in my investigation would have to be to see how the Society developed over its early years, just how it managed to establish itself as such an important force in the development of science and how it grew away from its Freemasonic roots.

CHAPTER 10

Early Works

It is clear that the Society had a kind of internal political history, not just of 'scientists' against 'amateurs' but among active enthusiasts with different interests and priorities, which sometimes surface but which is normally poorly documented.[1]

Michael Hunter, 1982

TO SURVIVE AT THE FOREFRONT of modern science for nearly three hundred and fifty years is no small achievement and to do so the Royal Society had to be special. I had already seen it grow out of a group of men who banded together in order 'to approach the knowledge of God through study of his works' and made their special concern the study of 'the hidden mysteries of nature and science'. However, I could not simply assume that its early history was just a harmonious blending of Roundheads and Cavaliers, responding to the benign guidance of Freemason Sir Robert Moray; that would be to greatly oversimplify the state of affairs.

I was able to establish a great deal about the formation of the Royal Society because its record-keeping is little short of amazing. Even such trivia as lists and memoranda survive from the seventeenth century and the Journal Books contain minutes right through from that very first November meeting in 1660. But, as I had noticed, there was an intricate web of relationships underlying the public image of the early set-up, and the 'Invisible College' which enabled Sir Robert to bring both sides together, so soon after the Restoration, was at the time a secret society. Why is there no mention of Freemasonry and its influence in these writings? The most likely reason is to be found in the vow of secrecy which every Freemason of the time had to take. This forbids the transmission of the secrets of the Order.

Any Freemason who joined the Royal Society would have repeated this solemn vow:

> *I further solemnly promise that I will not write these secrets, print, carve, or suffer them be done so by others, if in my power to prevent it, on any thing movable or immovable under the canopy of heaven, whereby or whereon any letter, character, or figure, or the least trace of a letter, character or figure may become legible or intelligible to myself, or to any one in the world, so that our secrets, arts and hidden mysteries, may improperly become known through my unworthiness. These several points I solemnly swear to observe, without evasion, equivocation, or mental reservation of any kind, under no less a penalty than to have my throat cut across, my tongue torn out by the root and my body buried in the rough sands of the sea at the low water mark, or a cable's length from the shore, whence the tide regularly flows twice in the course of a natural day.*

This injunction is quite clear and, when such penalties may not have been as symbolic as they are today, it tended to discourage loose talk. Moray was well known as a Freemason to many of his associates. He openly discusses the importance of his Masonry with Brother Mason Alexander Bruce in many letters, but at no time does he commit to writing any of the secrets of the Craft. At that time, it would have been considered secret knowledge that the study of the hidden mysteries of nature and science was the main purpose of the Fellowcraft degree. So perhaps the lack of written allusions to Freemasonry and its secret objectives is not surprising.

Freemasons may occasionally allude to its ritual in their writings and public exchanges, but they do so in a way which is only obvious to a fellow Freemason. In general, keeping to the letter of their vows, the Freemasonic links of some of these founders are just not written down at all. However, I have already drawn attention to many instances of the use of Masonic symbols and coded words within the group who founded the Royal Society.

Occasionally an extra layer of inter-linking between certain members shows. John Evelyn wrote to Bishop John Fell about the workings of the Council, 'Suffrages carry it, and not always the best Arguments fitter to be whisper'd in your Eare, than convey'd by a letter'. But he did not elaborate any further, in writing. When I reread William Preston's words I found that he says that many courtiers of Charles II became Freemasons. This was exactly what I discovered to have happened in the Scottish Court of James VI. Preston says that:

> *Some lodges in the reign of Charles II were constituted by leave of that noble Grand Master, and many gentlemen and famous scholars requested at that time to be admitted of the fraternity.*[2]

Was John Evelyn one of these gentlemen and famous scholars who was admitted to the fraternity? He was certainly in a position to join and I knew he had recorded a strange remark of the King's to Sir Robert Moray, concerning Joshua and the Valley of Joshaphat. Had that piece of Masonspeak struck a chord with him too! A letter which Moray wrote to Evelyn during the plague year of 1665 gave me a clue.

A Clue from the Black Death

Bubonic plague, formerly known as the Black Death, started in London in May 1665 and by July had reached epidemic proportions. By 12 August Pepys was writing in his diary:

> *The people die so, that now it seems they are fain to carry the dead to be buried by daylight, the nights not sufficing to do it in.*[3]

The king and Parliament moved out of the city, leaving the thoroughfare of Whitehall to become overgrown with grass. The death toll in London was of the order of one hundred thousand.[4] The king moved his court and mistresses to Oxford (he had another child by Barbara Palmer, now Duchess of Castlemaine, while he was there). Parliament was suspended and the Exchequer moved to Ewell. Sir Robert Moray and John Evelyn both left the city and while they were out of town conducted a correspondence which I found Masonically curious. On 14 January 1666 Sir Robert Moray wrote to John Evelyn:

> *By what telescope you read me at this distance I do not know. It seems you conclude me to be a greater Master in another sort of philosophy than that which is the businese of the Royall Society.*[5]

This letter, which Sir Robert Moray sent to John Evelyn, early in 1666, hints at a shared secret. Could that secret have been a shared knowledge of Freemasonry? I have already speculated that Evelyn might have been a Freemason. He had certainly dropped a few hints in his diary. But here was a letter from Moray to Evelyn, which seems to be using Masonic language.

When Moray wrote this letter he was staying with Charles's entourage, at Hampton Court. They had moved out of London to escape the latest outbreak of plague. Evelyn was in residence with the Lords of the Exchequer at Nonesuch Palace, Greenwich. He, also, had moved out of London to avoid catching the Black Death.

At first reading, this letter suggests a rather playful exchange between the two men. Moray accuses Evelyn of reading his secrets through a telescope. Normally a telescope is used to study stars. So Moray is referring to the star-shaped Mason's Mark he took as his own at his initiation into Freemasonry.[6] His casual allusion to the star, showing him to be a Master of another philosophy, has Masonic significance. Moray is using Masonic allusion and he expects Evelyn to understand it.

The Five-pointed Star

Fortunately, Sir Robert explained how he understood the five-pointed star when he wrote about its symbolism in a letter to Brother Mason, Alexander Bruce. In 1658 he said:

> *Astronomers might classify a starre as being of the least magnitude worth giving a name to, yet seeing, according to one of their maximes, the Magnitudes of them are not to be estimated by their appearances, but their situation and distance, those starres that are highest being sure to seem to be the least, though it is lyke enough they are the biggest: and the several distances is the ordinary reason given for the different appearances of magnitude.*[7]

After 1641 Moray always ended his signature with a five-pointed star. He interpreted his star symbol to say that what was important was what he really was, not what he appeared to be. He explained this saying, 'I had rather be somewhat of true worth though unknown, than appear to be what I am not, whatsoever the advantage of it.'

Moray also associated his Mason's Mark with secrecy. In the summer of 1667 he undertook a spying trip to Scotland on behalf of Charles II. Stevenson says of this trip:

> *Though in his later years Moray sought to avoid involvement in politics, in 1667 he agreed to visit Scotland and report to Charles II and Lauderdale, his secretary for Scotland, on conditions there.*[8]

Lauderdale was a Freemason, and Moray used Masonic language when reminding Lauderdale that parts of his reports would be written in invisible ink, beyond the open part of the letter, and that the start of the secret section would be after his Mason's Mark. In a report dated 1 July 1667 Moray wrote:

> *Wher you see my Mason Mark you will remember what it meanes. I will play the Mason in my next.*[9]

On 15 July he again reminded Lauderdale in a letter:

> . . . *the next time you converse with the starres, you will get the Gleanings of our discoveries.*[10]

The five-pointed star was evidently a very important symbol to Sir Robert and he associated it with secrecy, invisibility and a symbol that stood for true worth. In these two examples of his correspondence with fellow Masons he assumes they will understand the symbolism of the pentangle and he has good reason to do so because this symbol is more than just a personal identification mark for Sir Robert. It is also one of the most important symbols used in Freemasonry.

Many modern Freemasons' lodges in England still have in the centre of the lodge a five-pointed star containing the letter

G and lit from behind with a bright light during some of the rituals. In particular this symbol is lit during the second degree when the new Fellowcraft is told that the combined symbol is known as 'the Glory' and that it represents God, the Grand Geometrician of the Universe. The purpose of the stars is described in the Second Degree ceremony:

> *Besides the Sun and the Moon, the Almighty was pleased to bespangle the ethereal concave with a multitude of Stars, that man, whom He made, might contemplate thereon, and justly admire the majesty and glory of his Creator.*

Stevenson says of Moray's use of the five-pointed star:

> *the suitability of the star/pentangle as a mason mark is revealed [in Moray's correspondence]. Masons were not what they seemed, in that outsiders could not see anything distinctive about them which identified them as Masons, but fellow initiates could detect 'invisible' emanations which identified them. As with the stars, there was more to Masons than appeared at first sight.*[11]

Even this, however, is not the full story. Sir Robert was descended from Freskin MacOleg of Moray and another branch of this family became the Earls of Stormount. Sir David Moray was made first Viscount Stormount and Lord Scoon when James VI gave him Scoon Palace in 1604. The Murray family, as they now spell their surname, still own Scone Palace (as they now spell the name of their house), and when I had visited the place, researching the circumstances of the coronation of Charles II, I had noticed the extensive use of the five-pointed star throughout the Palace. The ceilings of the Library and Ambassador's room could easily pass for the ceiling of a Freemasons' Lodge. Into each of the ceilings is let a gold pentangle with the main room light hanging from its centre. The small chapel on Moot Hill has dozens of star symbols engraved in its walls. Evidently the star symbol had a family connection as well as a Masonic connection for Sir Robert.

The importance of this symbol to Moray makes his letter to John Evelyn even more powerful. He first drops the broad Masonspeak hint that Evelyn is viewing him as 'starre' by suggesting Evelyn must be using a telescope to see him clearly at a distance and then he goes on to 'play the Mason' by hinting that he and Evelyn are Masters (Right Worshipful Masters, perhaps?) of another philosophy which underlies the Royal Society.

Evelyn also started to use the star mark with his own signature early in 1660. It is quite permissible for more than one Freemason to take the same mark, provided they are not members of the same lodge at the same time. When I became a Mark Mason I also took the five-pointed star as my Mason's Mark. Did Evelyn take the five-pointed star as his Mason's Mark too? He seems to have been made aware of the formation of the Royal Society at a very early stage, the stage when Moray seemed to be involving as many Freemasons as possible. On 6 January 1661, five weeks after the initial meeting, he first mentioned the Royal Society in his diary:

> *I was now chosen and nominated by his Majestie, for one of the Council by sufferage of the rest of the Members, a Fellow of the Philosophical Society, now meeting at Gressham Coll: where was an assembly of diverse learned Gent: It being the first meeting since the returne of his Majestie to Lond: but begun some years before at Oxford and here in London: during the rebellion.*

There is no record of him having taken any interest in either

Wilkins' or Ward's meetings, and yet he seems to have been aware of them. If the meetings were lodge meetings and Evelyn had become a Freemason, then there is no mystery. If he was not a Freemason, then he seems remarkably informed about private meetings of Senior Parliamentarians, and considering his own Royalist connections this is unlikely. Nor could he have learned about them from Sir Robert Moray as he did not meet him until 9 March 1661. Evelyn says quite clearly that he was recommended for the Society by Charles II. Perhaps William Preston is correct when he says Charles II was a Freemason. If both the king and Evelyn were Freemasons and were aware of Moray's plan to form a scientific society using his Masonic contacts, it would certainly explain why the king proposed Evelyn for the Society as soon as it was formed. It would also explain why Moray started to play Masonic word games with Evelyn about the hidden meaning of the star and the secret philosophy of science which they both seem to know lay behind the new Society. It could even explain why Evelyn then started to use the star as an addition to his own signature.

I had, however, one more clue which suggested that Evelyn might also have been 'Master in another sort of philosophy than that which is the businese of the Royall Society'. That clue was hidden in notes I had made many years previously.

A Trip to Wotton House

John Evelyn was born at Wotton House near Dorking in Surrey. He was the younger son of Richard Evelyn. John was eventually buried at Wotton House in 1706, having inherited the property from his brother. In the 1980s Wotton House was leased from the Evelyn family, by the Home Office, and used as a Fire Service Staff College. At this time I was often invited to lecture on the residential Brigade Command Courses which were run at the College for senior fire officers and I would stay in the house for the duration of the courses. The first time I went there I was greeted by the staff officer, Senior Divisional Officer Ron Shettle, whose responsibility it was to look after visitors. Ron was very interested in the history of the house and in the Evelyn family. He was keen to show me around. I asked if I could see the room where John Evelyn had been born. That room, which looked out over the lawns to the front of the house was then the typing pool. The gardens to the front of the house had been landscaped by Evelyn and the most striking feature of his work was a large fountain. Evelyn had piped a number of streams into a leat, which channelled the water beneath a pillared Temple into a great pipe. The head of water was enough to give a fountain jet which sprayed a good five metres into the air. As we walked back along the terrace Ron pointed out a very mature tree which was covered in red berries.

'That's a mulberry tree,' he said. 'John Evelyn brought it as a cutting from his house at Sayes Court and planted it here when he inherited the house.'

Ron went on to tell me that William Oughtred had visited Wotton House in 1655, although at that time the name had meant little to me. With the benefit of hindsight I now realised that Oughtred had been Ashmole's patron when he arrived in London and, as I have already mentioned, seems to have been a Freemason. Did Oughtred introduce Evelyn to Freemasonry, among the 'other discourse' Evelyn mentioned, but did not record, in his diary for that day on 28 September?

I had asked Ron where John Evelyn was buried and he had told me that Evelyn's grave was in the Family Vault at Wotton Church. Wotton Church is a fascinating building. It has an unusual tower and still retains the colourful Catholic decoration which Oliver Cromwell's soldiers usually destroyed. For some unknown reason Cromwell did not allow his troops to damage the Evelyn Family Chapel. Cromwell had also spared the Freemasonic shrine of Rosslyn and I had once been told by a vicar of Rosslyn that Cromwell had spared the Chapel because he was a Freemason and recognised the symbolism in the building. As Oliver had also spared Wotton Church, had it been because Evelyn's family was known to be connected with Freemasonry? As I reviewed my old diary entries I couldn't help remembering Oliver's unusually sympathetic treatment of Inigo Jones.

When I had visited Wotton Church I had made a note of the epitaph which marked Evelyn's tomb and reading it again a certain line stood out:

> *& perpetuated his fame by far more lasting Monuments that those of Stone or Brass.*

I couldn't help thinking of one of the versions of the Masonic story of Enoch. Enoch, realising that a great flood was coming expected to die. He carved the secrets of science on two pillars. One was made of stone and one of brass. In one ritual it is said:

> *Then, fearing that all knowledge of the arts and sciences would be lost in the universal flood, he [Enoch] built two great columns upon a high hill – one of brass, to resist water, and one of granite, to resist fire. On the granite column was written in hieroglyphics a description of the subterranean apartments; on the one of brass, the rudiments of the arts and sciences.*[12]

Was Evelyn's epitaph a veiled reference to this well-known Masonic story? To a Mason it can mean that Evelyn knew he had left behind a greater legacy in the Royal Society than Enoch had done with his pillars of science.

On balance I think it is quite likely that John Evelyn was a Freemason, but no lodge records exist to show his initiation and he never made more than veiled references in his diary. That question of his membership of Freemasonry is unlikely ever to be fully answered, but if Evelyn was a Mason it helps explain his role in the Royal Society.

A Major Force in Science

An application of the fraternal bonds of Freemasonry, and the use of Freemasonic ideas, is not enough to explain the success of the Royal Society. It lists among its members some of the most important scientists of the last few hundred years. They range from Sir Isaac Newton to Lord Rutherford and Professor Stephen Hawking. So just how did the Society manage to create a climate that promoted such scientific creativity? It is certainly not anything modern Freemasonry can match. Freemasonry today is mainly ignored by the professional classes and young scientists. The Royal Society, however, is still a major force in science. Sir Henry Lyons says of it:

> *Since its foundation in 1662 the Royal Society has grown in size and influence until is it now widely recognised as an institute which is playing an important part promoting research in all branches of natural science.*

What made it into such a fantastically successful organisation? It borrowed a philosophy from early Freemasonry and turned it into a force which changed the very nature of the world. It did this by subverting and applying a concept which says that understanding comes from observation and experiment, not just from philosophy. The success of this approach is to be seen everywhere where science has led to technological advance. Without this basic change in attitude we would not have electronics, genetics, biochemistry or nuclear engineering. The Royal Society gave us the very stuff of rocket science that we now take so much for granted.

There are various distinct stages in the Society's early development. It starts with a formative period, prior to the issue of the First Charter, when Sir Robert Moray is clearly the driving force. It then moves into an active period, driven by a lively Council. As the interest of the early Council members flagged, or they died off, the Society went through a difficult period, when it looked as though it might fail. A new influx of active members under the presidency of Isaac Newton then opened a staid, but stable and scientifically successful, period that continued until the early nineteenth century.

By the mid-nineteenth century, after a period Lyons calls the Scientific Revolt, the Society became the important forum for scientific research and debate it is today. By then the importance of the 'amateur' members had passed and to become an FRS was now a mark of scientific distinction. I felt that I now needed to investigate the detail of the early period to see what had laid the foundations of this later success.

The rejection of the First Charter seemed to indicate a shift of power away from Moray. Up to that point he had been the most regular president, elected month after month. From the time of the First Charter, in 1662, Moray pushed Brouncker forward. By this time Sir Robert was in his mid-fifties and he may have wanted just to play a less energetic role in the Society, but his level of enthusiasm stayed the same, he simply directed it in a different direction.

Lyons said of Moray during this period:

> *From the time he returned to London in August 1660 he had devoted himself wholeheartedly to making the Society the important institution that in his opinion it should be. During 1661 and the first half of 1662 he was virtually President of the Society though the title was not formally used by anyone until the Charter had been sealed.*[13]

I decided to start by looking at just what Sir Robert did next and who he did it with.

Ships, Weapons and Navigation

One of the major problems faced by Charles II's Navy was navigation. There were three aspects to this problem. In descending order of difficulty these were:

1. The measurement of Longitude, which involved working out the difference in time between the reference point, (now set at Greenwich, but at that time at the port of origin) and the local time at the ship's position.

2. The Measurement of Latitude, which involved developing some way of accurately measuring the local time, and the height of the sun above the horizon at local noon.

3. The measurement of direction, which involved finding the direction of the earth's magnetic field and comparing it to astronomical sightings of true North.

From its earliest meetings Moray encouraged the Society to work on all three of these problems. This had prompted Laurence Rooke to suggest that the rotation of the moon could be used as a giant sundial, visible from anywhere on the earth. At this point Sir Robert had suggested to the newly formed Society that Christopher Wren create a working model of that lunar sundial so the method could be tested. It was that lunar model which eventually ended up in the archives of Charles II when Sir Robert realised the pendulum clock was a better way forward. However, the early work by Bruce and Holmes, which I discussed in Chapter Eight, showed that pendulums were too susceptible to the adverse motion of ships to keep reliable time. This left two possible solutions to the longitude problem. Either improve the accuracy of mechanical clocks or develop Rooke's idea of using some regularly occurring celestial events that could be observed from anywhere. Moray encouraged the

Society to actively pursue both these options.

While Moray had been in Paris, in 1661, there had been talk among Paris astronomers of using movements of the Moons of Jupiter to provide such an event. On his return to London Moray had spoken to Laurence Rooke (who was at the 28 November meeting), and had become interested in another of Rooke's ideas, that of using the relative visibility of Jupiter's or Saturn's moons to calculate longitude. Rooke's method, which was published posthumously in Spratt's *A History of the Royal Society*, suggested that the eclipse of a satellite occurred at the same instant of time in London and where the ship was. By noting the time of these events, the local time of the eclipse could be compared to the time of the eclipse computed for London and the difference in longitude found. Moray realised the importance of Rooke's idea and set about creating a means to implement it.

Before Rooke's suggestion could be turned into a practical proposition, tables of the times of transit of Jupiter's and Saturn's moons would need to be prepared. This could only be done with a powerful telescope and an accurate clock. Huygens had created just such an accurate observatory clock. To demonstrate the method of creating these tables Moray persuaded the king to set up a telescope at Whitehall and then, using the king as social bait, he invited influential members of the new, as yet unchartered Society to a demonstration of the method. John Evelyn attended the event and recorded it in his diary entry for 3 May 1661:

> *This evening I was with my L Brouncker, Sir Rob; Morray, Sir Pa: Neill, Monsieur [Huygens] & Mr Ball (all of them of our Society & excellent Mathematicians) to shew his Majestie (who was also present) Saturn's Ansatus as some thought, but as Monsieur [Huygens] affirmed with his Balteus (as that learned gent had published), very neere eclipsed by the Moone,*

> *neere the Mons Porphyritis; Also Jupiter & Satelites through the greate Telescope of his Majesties, drawing 35 foote; on which were divers discourses.*

Six days later Evelyn records that he and Moray met with 'Dr Wallis, Professor of Geometry in Oxford, where they had discourse of several Mathematical subjects'. Moray was preparing his ground well. The equipment and goodwill for experimental observation of Jupiter's moons was ready, now he needed someone to calculate how to draw up the necessary tables. Wallis, at that time was the world's expert on the motion of planets, being very close to developing the system of calculus that Newton finally perfected. Ball set about collecting the basic observations of Saturn while Rooke studied Jupiter. If Rooke had not died prematurely this initiative of Moray's might well have resulted in the Royal Society being the first to produce a full set of tables of the transits of Jupiter's moons. Rooke's death, however, put back the work and Guiseppe Campani of Rome won the race to publication in 1668. Ball's tables of transits for Saturn, however, were the basis for John Wallis's work on cosmology.

At the time of the first meeting in November 1660, Robert Boyle had a very able technician working for him at Oxford. It was this man, Robert Hooke, who became an important scientist in his own right and was appointed the first Curator of Experiments for the Royal Society. Hooke discovered a law of physics which describes how the behaviour of a spring can be predicted. He carried out a series of experiments carefully measuring how far different springs stretched as he hung various weights on them. He discovered that 'the force tending to restore a spring to its equilibrium position is proportional to the distance by which it is displaced from the equilibrium position'. This law, known as Hooke's Law, had one very practical application. Hooke noticed that a spiral spring will expand and contract about its rest position in equal periods of time, no matter how far the spring moves, provided it does not distort. This is the basic idea behind the pocket watch. Eventually Hooke would realise that he could replace the bulky pendulum with a small spring-loaded balance wheel, which was potentially more accurate. More importantly for a marine clock, the spring balance wheel would still keep good time even when it was tilted sideways.

In 1665 Moray and Bruce were attempting to get the king to grant a patent for a Marine Pendulum Watch which they had developed with Huygens. Hooke, as curator of Experiments for the Royal Society, was asked by Brouncker to carry out further sea trials of this pendulum watch on various naval vessels. Hooke was unimpressed, writing in his journal:

> *There is no certainty of measurement to be had from pendulum watches for the determination of longitude because*
>
> 1. *They are never hung perpendicular, and consequently the checks are false.*
>
> 2. *All kinds of motion upwards and downward will alter the vibrations of them.*
>
> 3. *Any lateral motion will produce yet a greater alteration.*

Now, at Moray's request, Hooke, as curator of experiments, had carried out a series of tests on the working of pendulums so his opinion was firmly based in experiment. On 25 August 1664 Hooke wrote to Robert Boyle describing pendulum experiments he carried out at Old St Paul's in London: Hooke climbed to the top of the rather ramshackle steeple, which had been struck by lightning, and swung a long pendulum from the top. To time its movement he had with him a smaller pendulum which had a half-second period, as this was before he had invented the pocket watch! Here is what he wrote to Boyle about the experiment:

> *A pendulum of the length of one hundred and eighty foot did perform each single vibration in no less time than six whole seconds; so that in a turn and return of the pendulum the half-second pendulum was several times observed to give twenty-four strokes or vibrations. Another was, that this long pendulum would sometimes vibrate strangely, which was thus; The greatest part of the line, by guess about six score foot of the upper part of it, would hang directly perpendicular, and only the lower part vibrate; at what time the vibrations would be much quicker, and this though there was a weight of lead hung at the end of the string above four pounds weight.*[14]

The tower had been quite badly damaged by the lightning strike so that all its inner floors were destroyed. Although this made for a useful experimental space, two hundred and four foot high, it could not have been a very safe working environment. Hooke commented to Boyle that the steeple was 'without any kind of lofts but having only here and there some rotten pieces of timber lying across it'. Hooke had some minor accidents while trying to haul rather delicate barometers up and down the tower to measure the changes in air pressure. He complained: 'the glass being but thin, was broken'.

Once he had worked out how to haul his instruments up and down the tower without wrecking them he invited Moray and Brouncker to observe. Lisa Jardine comments:

> *[Hooke] then carried out more pendulum experiments at Old St Paul's, this time with the President of the Royal Society (Lord*

Brouncker) and Sir Robert Moray in attendance (presumably safely at the bottom of the structurally unsafe steeple).[15]

As Hooke learned more about the shortcomings of pendulums as timekeepers he grew to understand that his work on the behaviour of springs could be used to improve on Huygen's pendulum clock. Just six months later, as Moray achieved the Royal patent for pendulum watches on behalf of the Royal Society, Hooke announced to Brouncker, Wilkins and Moray that he had invented a better type of marine watch. Hooke's watch made use of his knowledge of the compression and extension of a spring and was the forerunner of today's balance wheel watch. He asked for the Society's support to carry out sea trials and was given £10.[16]

Moray mentioned, when describing Hooke's invention in a letter to Huygens, that he considered Hooke's disclosure of the secret of the spring balance clock to the Society was an act of great generosity. It was an act that Hooke would later regret as in 1675, two years after Moray's death, Huygens tried to patent an almost identical spring-controlled watch. The result was a bitter quarrel between Hooke and Huygens. But Hooke had no patent to prove his prior claim. As he had been about to lodge it with the king, the plague had hit London and everyone had moved away from town. Hooke went with John Wilkins to Epsom, and while there created an improved method of taking a sighting of the height of the sun, (the second part of the navigation problem). Hooke and Wilkins spent the duration of the plague playing with reflectors and dashing about the grounds of Durdans House in Wilkins' new carriage, to test the accuracy of Hooke's travelling clocks.

The second component of the problem of navigation involved finding latitude, which is done by measuring the height of the sun above the horizon at noon (local time). This had traditionally been carried out using a sighting stick, basically a rod with a cross piece held at arm's length. Hooke invented an instrument for accurately focusing the sun's rays onto a scale using a reflecting quadrant. The same instrument could also be used by a surveyor to measure the height of buildings, by sighting on the top. Instruments of this type are in use today in the form of the navigator's sextant and the surveyor's theodolite.

Hooke wrote to Robert Boyle on 15 August 1665 to say that he had taken Bishop Wilkins to Old St Paul's steeple and tested his reflecting quadrant. Hooke had previously measured the exact height of the steeple during his perilous pendulum experiments. He reported he had used his new device to measure the height of the steeple to an accuracy of 12 foot in 204 (i.e. better than 6 per cent error), 'which is more than is possible to be done by the most accurate instrument, or the most exact way of measuring'. Hooke intended to demonstrate his new reflecting quadrant to the Royal Society on 12 September 1666, when he also planned to report on his surveying of the St Paul's Steeple with a view to carrying out repairs. He was, however, overtaken by events when St Paul's burned down, in the Great Fire of London, a week before the meeting. Hooke did not get to present his findings as the meeting was cancelled.

According to Jardine, Hooke also 'worked on methods and instruments for finding local time at sea' (the crucial calculation needed alongside the time at home 'held' by the longitude timekeeper)[17]. The Royal Society was focusing on all aspects of the navy's navigational problem.

Moray had even made sure that accurate use of the compass for direction finding was not overlooked. As improved methods of astronomical determination of ground-based latitude and longitude were developed it was found that the boundary lines drawn by use of the magnetic compass were in error. On those occasions when these errors related to the boundaries between colonies in the New World, some owned by Britain, and adjoining ones by the Dutch who were at war with England, serious problems arose.

The marine compass had been known in England since 1180, when Alexander Neckam, who had studied in Paris, wrote the first book about its use in navigation. Neckam had believed that the compass pointed to the North star, driven by some mysterious astrological power, but in 1600 Queen Elizabeth's physician, William Gilbert, had shown that the Earth itself was a giant magnet, but that its magnetic pole was not in the same place as the true North pole. In the intervening years sailors had noticed that the angle of the compass needle to the horizontal, called the declination of the magnetic field, varied with latitude. Moray saw another opportunity to improve navigation and persuaded William Ball to carry out measurements of the direction and declination of the Earth's magnetic field over a number of years. Indeed, it was Ball's monopoly of the equipment in order to carry out routine collection of magnetic information that caused the Council to suspect that Ball had absconded with the Society's 'Magnetical Equipment'. This was when Sir Robert had had to speak up for him, eventually purchasing the equipment on Ball's behalf so he might carry on with his work.

It is not clear how much even some of the better scientists, such as Robert Boyle, really understood about magnetism since Boyle had written to Hooke, while the latter was risking life and limb swinging pendulums from the top of the semi-derelict St Paul's steeple, suggesting that Hooke might also see if he could measure any difference in magnetic declination between the top and the bottom of the tower. Hooke wrote back saying the experiment was pointless as the tower was bound together with rusting iron bands.

Meanwhile, Ball was experimenting with larger and larger compass needles, the largest he built was over twenty feet long.

He wanted to know if size mattered and if longer needles gave greater accuracy. He did find that the magnetic north pole seemed to move very slowly over a period of years and so laid the foundations of the tremendous survey work on magnetic declination which Edmund Halley would carry out on behalf of the Royal Society in the 1700s.

All in all Moray made sure that all aspects of navigation figured in the very first experimental schedules of the Society. He may not have held the title of President but he seemed to have been pulling all the strings.

Brouncker was the first chairman to use the title President and not only was he extremely interested in ships and naval matters, he was also about to be appointed a commissioner for the navy. So, with Brouncker in charge, Moray could be sure that the Society would pursue matters of naval interest.

This may not be the whole story, however, as another reason is also indicated by the choice of Brouncker. Brouncker had stayed in England during the period of the Protectorate but had kept a low profile. He had been elected a member of the Convention Parliament of 1660 and while only nominally a supporter of Parliament, he had not been part of the exiled opposition. In other words he was a safe pair of political hands in which to place the Society and to keep it focused on naval problems in its early years. There was clearly still strong feeling in the Society that would not allow too close a relationship with the Monarchy and it was during this period between the first two charters that a considerable amount of underground manoeuvring took place to settle this matter.

Once the matter of title and relationship to the king had been settled, however, the Society quickly started to develop in ways which must have greatly surprised Sir Robert. It did certainly address many important naval problems, in fact for its first three years it seemed to carry out the main responsibilities of the Navy Board in respect of scientific problems, but it also became extremely fashionable. This resulted in an influx of amateur members who were mainly interested in using the meetings as a source of entertainment, but they did contribute useful amounts of money. The Society was very open in its membership but its fees were high. As historian Michael Hunter comments, 'Members of the aristocracy were positively encouraged to join, and privy councillors and those above the rank of Baron were admitted without scrutiny.'[18] Clearly it helped to live in London and have money if you wanted to be a Fellow during this fashionable period, although useful scientists of limited means often had their subscriptions waived. (An example of this happening is John Collins (Fellow Number 235) who was accepted without fees purely for his scientific skill.)

The main direction of the Society was, however, now under the control of the Council and it was made up of a number of men who, although not necessarily capable scientists, were interested in experiment for its own worth. Because the scientists in the new Royal Society were outnumbered two to one by the amateur gentlemen members, the Council needed to spread its scientific talent about if it wanted to try to carry out the objectives Charles had set it in his Second Charter (i.e. to extend the boundaries of the Empire and the very arts and sciences).

Brouncker was riding high at this time; he had been appointed President of Gresham College. The Second Charter had confirmed him as President of the Royal Society. Then in December 1664, the king made him an extra Commissioner to the Lord High Admiral, James, Duke of York. In his new government post he would have been aware of the new demands that would soon be made on the run-down navy. In 1663 Charles had granted a patent of proprietorship to a company of eight of his courtiers, including Edward Hyde, the Earl of Clarendon, to occupy the coast of America from

Virginia down to the Spanish Colony of Florida. (This land would become North and South Carolina.) This expansionism left a much longer coastline to defend against the Dutch who had a strong fleet trading with North America and a naval base at New Amsterdam (now New York). He would also have been aware of the deteriorating relations with the Dutch, which had recently been inflamed by the 1661 Navigation Act, repeating Cromwell's provision that foreign merchant ships were not to carry cargo from English colonies but adding that all English Colonies could only export certain raw materials to England, and nowhere else! Soon after Brouncker was appointed as President of the Royal Society Charles passed another Act, this time for the Encouragement of Trade. This said that all European goods destined for the English Colonies must first be brought to England, unloaded and thence shipped overseas in English-built ships.

Since the First Dutch War a great rivalry in both trade and fishing had developed between the English and Dutch fleets. Now the Dutch were winning the trade war. The Act of Navigation and the Act for the Encouragement of Trade were attempts to legislate a victory for English shipping which had not been possible by free competition and it was not working. Popular opinion was pushing the newly restored king to do something to stop the Dutch stealing not only British herrings but the long-distance trade routes as well.

Charles's Parliament, whipped up by its merchant class members, was strongly in tune with the public dislike of the Dutch. In 1663 they voted the king enough money to build and equip over one hundred new ships, and fit them with new and heavier cannon.[19] Now it was clear why Charles insisted that Brouncker, with his interest in innovative shipbuilding, navigation and naval weaponry, had been put in charge of the Royal Society. In its early years the Royal Society was the *de facto* technical support department for the Office of the Naval

Commissioners. Parliamentary money without technical expertise would not have improved the Navy. Moray's tactics were beginning to pay off.

Historian J F Scott had noticed this link when he said of the early days of the Restoration:

> *It is intriguing to look back on the close connection of the Royal Society with the Royal Navy in those critical years, when a Royal Fellow, two Presidents and several Fellows were involved in its administration.*[20]

The Royal Fellow was Lord High Admiral, James, Duke of York, Fellow 181, admitted in January 1665 at the beginning of the Second Dutch War. The two Presidents were Brouncker, and Samuel Pepys, Fellow 187, admitted a month after James. The other fellows were Peter Pett, Fellow 112, admitted September 1662 and Matthew Wren, Secretary to the Lord High Admiral, who had been on the list of January 1661 and was Fellow 21. The Fellows had formed a majority on the Board of the Naval Commissioners so I was interested to see if there was any evidence that they had known each other before they came to work together. I soon found that there was!

Just before the issue of the First Charter, William Petty's new twin-hulled sloop *The Experiment* beat the king's mail boat in a race from Holyhead to Dublin. Petty's work on ship design introduced many ideas for construction which would be taken up by the English shipbuilders during the surge in shipbuilding the Second Dutch War provoked.[21] *The Experiment* was built by Peter Pett who, with his brother Christopher, was the most productive shipbuilder of the time. The Petts' shipyards were in Deptford. Christopher built racing yachts for the Duke of York, and the Petts led the way in introducing these light and compact vessels, which could sail very close to the wind. Skill in the design of extremely manoeuvrable sailing vessels was to prove very useful to the Navy Board as they re-equipped the fleet after the Restoration.

Charles had encountered his first yacht in 1659 when he had travelled from Breda to Delft on *The Mary*, a hundred-ton trading vessel belonging to the Dutch East India Company. Brouncker had taken the design ideas of this little ship and in company with Kenelm Digby and William Petty had commissioned an improved pleasure vessel for the king. A ship for which Peter Pett set himself the target 'to outdo this (the design of *The Mary*) for the honour of his country'.[22] The ship was called *The Greyhound*, because of its speed and manoeuvrablity. Peter Pett had built the design, at a cost of £1,335, and Brouncker presented it to the king in 1662. (The king later indulged himself in fitting out the main cabin with a crimson and damask bed and hangings decorated with gilt leather. After all it was intended to be the king's pleasure craft and needed to be properly equipped for the king's favourite pleasures!)

Another of the fellows on the January 1661 list had been Sir Peter Pett, who was a first cousin to shipbuilders Peter and Christopher. Although Sir Peter was a lawyer he preserved the family interest in ship design. On 17 September 1662 Brouncker had admitted Sir Peter into the Royal Society, on the recommendation of Robert Boyle. Boyle at the time had been working on a diving bell which was being built in the Pett's family shipyard. He must have been impressed with Peter's knowledge of the practicalities of shipbuilding and so invited him into the society to strengthen its ability to study naval problems. Sir Peter had been a naval commissioner under Cromwell and had been kept in post by Charles. So by 1664 the Board of Naval Commissioners was practically a sub-committee of the Royal Society.

The technical support of these members of the Royal Society was the main reason that by June 1665 the Duke of York was in command of a fleet of 150 ships, manned by 25,000 men and mounting 5,000 cannon, to face the Dutch fleet at the Battle of Lowestoft. James managed to drive off the Dutch but did not defeat them. At first the war went well, New Amsterdam was taken and the English Fleet took many Dutch merchant prizes in the narrow seas of the Channel. But, the Second Dutch War was far from over.

Meanwhile the Society became a centre for the transfer of information. Its freedom, confirmed in the Second Charter, which Sir Robert Moray had negotiated, allowed it to correspond with foreign scientists. Sir Robert had gained much of his own scientific knowledge by corresponding with leading experts during his exile. He believed in sharing information and insights so the range of the Society's correspondence developed rapidly.

Philosophical Transactions

The idea for a regular scientific journal came originally from Sir Robert Moray. In September 1661 he wrote to Huygens that he intended that as soon as the Society had received its permission to publish that 'we shall print what passes among ourselves, at least everything that may be published. Then you shall have copies among the first, and if there is something withheld from publication, it will be much easier for me to communicate it to you than to have to send word of everything by letter.'[23] As we shall see, within four years this idea would eventually develop into the world's very first scientific journal.

The Second Charter confirmed that the Society should have two secretaries and the king chose the first incumbents, 'the aforesaid John Wilkins and Henry Oldenburg to be and become the first and present Secretaries of the aforesaid Royal Society: to be continued in the same offices until the aforesaid feast of St Andrew the Apostle'. John Wilkins is a familiar name, and it is clear that by now the king had fully forgiven him for marrying a Cromwell, but who was Henry Oldenburg?

Oldenburg had figured in the list of people invited to join, which had been drawn up immediately after the first meeting, and he is listed as Mr Henry Oldenburg, Literary.[24] The Royal Society has a portrait of him, painted by John van Cleef. He appears a rather chubby man with long curling locks, a centre parting and a pencil line moustache. He is dressed in a black gown with a white bib collar and ornately embroidered cuffs. In his right hand he holds an open pocket watch and his left hand is clutching his breast with his thumb extended at right angles. The pose is strange, as it appears at first sight that he has adopted the posture a Freemason would take up if he was making the sign of fidelity, but a closer examination shows that his pose is actually a mirror image of this sign. It may well be coincidence but the effect is uncanny. As I looked at the picture I made the sign of fidelity and the portrait seemed to be reflecting my posture.

Oldenburg has a slightly unkempt look in this portrait as if he had been caught by a camera without time to comb his hair. Compared with portraits of the other founders Oldenburg looks like the poor relation, and perhaps he was. He was born in Bremen around 1615, making him forty-five when the first meeting of the society was held. Bremen was a small independent state to the northwest of modern Germany. Oldenburg first came to England at the age of twenty-four as a new graduate Master of Theology, earning his living as a tutor to the nobility. He left England just before Charles I was defeated, rumours said he had been a supporter of the king and was afraid of retribution from Parliament. Whatever his reasons for leaving he spent the next few years travelling around the continent of Europe. In 1652 he returned to Bremen. He had only been back a year when he was appointed diplomatic agent for the Senate of Bremen with the brief to attend the Court of Lord Protector Cromwell and make sure that Bremen stayed neutral during the First Dutch War. Historian R K Bluhm says of this appointment:

> *His appointment to this post was criticized on the ground that during his former stay in England he had taken the King's side in the quarrel with Parliament, but whatever truth there may have been in this assertion his experience of the country and his knowledge of the language were of more importance in the Senate's eyes, and on 30 June 1653 he received his letter of appointment.*[25]

He must have done his job reasonably well because at the end of the First Dutch War he was kept on and stayed in London. This job lasted until around 1655 when Oldenburg once more took a job as a tutor. This time it was to Richard Jones, who was the nephew of Robert Boyle. Oldenburg went with Richard Jones to Oxford, where they both enrolled as 'strangers' to study at the university. During his time at Oxford, Oldenburg struck up an acquaintance with Boyle who invited him to some of the meetings that Wilkins hosted at Wadham. (Perhaps his choice of a Masonic style pose for his official portrait is not entirely accidental if he joined this group which may have been an offshoot of a Masonic lodge.) Oldenburg then took his young pupil travelling in Europe during the period leading up to the Restoration and did not return to England until after the king's return was confirmed.

He spent the final year before the Restoration in Paris where he was a regular visitor to the house of the French historian M. De Thou. De Thou had founded a literary society which also attracted some men of science, as well as politicians. The meetings of this group served as 'bureau for the exchange of foreign news' and was at times very political in its interests.[26] This may well have been where Oldenburg was first exposed to the idea of using a correspondence circle as a means of gathering sensitive information with a market value.

Oldenburg's appointment as a secretary alongside John Wilkins is another apparently arbitrary choice by the king which, like that of Brouncker, has puzzled historians. Bluhm says of it:

> *On whatever grounds the appointment was made, it proved to be an excellent one, and Oldenburg henceforward devoted himself to the Society's affairs with great industry.*[27]

But perhaps the king's decision was not so arbitrary! Oldenburg had a flexible approach to political affiliations and had a tendency to seek out favour wherever possible. He had used his political links to Charles I to get himself appointed an ambassador to Cromwell's Court. There he made friends with Cromwell's Latin Secretary, John Milton, and it was through Milton he got the job as tutor to Richard Jones. He then used Jones to become friendly with Robert Boyle and through him was invited to the Oxford group meetings. If there was a lodge connection with these meetings, as Wallis's evidence seems to suggest, then it is quite in character for him to have joined any lodge which might help his ambition.

In the Second Charter Oldenburg's name is closely linked with John Wilkins, who all the contemporary writers agree, was the leader of the Oxford group. Oldenburg seems to have been good at currying favour and quick to spot opportunities for his own advancement.

When Moray put forward the idea of creating correspondence circles within the Society, and publishing the results of experiments, Oldenburg saw his chance to create a small side business in other types of information. He mentions this in a letter he wrote to Robert Boyle, soon after he had become one of the secretaries. This is what he said:

> *Sir, give me leave to entreat you, that in case you should meet with any curious persons, that would be willing to receive weekly intelligence both of state and literary news, you would do me the favour of engaging them to me for it. The expenses cannot be considerable to persons that have but a mediocrity; ten pounds a year will be the most, that will be expected; eight or six pounds will also do the business.*[28]

To be fair to Oldenburg he needed some sidelines if he was to earn enough money to survive. He was not a wealthy man and the post of Secretary to the Royal Society only paid £100 a year, if the Society could afford to pay, which it did not always manage to do. To be able to sell on the political gossip he collected from the Society's overseas letters must have seemed like a good perk of the job for him.

Another important sideline he developed was to write and publish the journal which Sir Robert Moray had suggested. Oldenburg edited and reported the experimental work of others, and in doing so established the first tradition of independent review before experimental results are published. He took the reports of the experimenters and edited the work into readable articles. This tradition has survived to the present day, developing into the system of refereeing articles for academic journals.

It is fortunate that Henry Oldenburg needed to earn additional money. If he had been independently wealthy he might never have been encouraged to develop Sir Robert's idea of a regular printed journal.

Oldenburg would never have been allowed to publish any sort of journal, or conduct an international correspondence circle without the support of the Royal Society. In the 1660s nobody was allowed to publish anything unless they had a special charter from the king to do so. Oldenburg was allowed free use of the Society's right to publish. This freedom to print and sell must have been a tremendous incentive to Oldenburg to act on Sir Robert's suggestion and create a regular record of experiments.

On 1 March 1665 Sir Robert had put a motion to the Council which said that:

> *The Philosophical Transactions, to be composed by Mr Oldenburg, be printed the first Monday in every month, if he have sufficient matter for it, and that the tract be licenced . . . and that the President be now desired to licence the first papers thereof.*[29]

Within a week of the Council accepting this motion, the first peer-reviewed scientific journal in the world was published. The title was snappily phrased for the time – *Philosophical Transactions: giving some account of the present undertakings, studies, labours of the ingenious in considerable parts of the world.* This first edition consisted of sixteen quarto pages and was written, printed and produced at the personal expense of Henry Oldenburg, although published under the imprimatur of the Royal Society. Once printed, Oldenburg was free to sell his journal with a total monopoly on scientific publishing.

The importance of this new venture to the development of science was enormous. Modern science, we have seen, develops by a process of observation, prediction and finally control. The idea of a journal, recording and spreading the detail of observations and allowing the sharing of predictions, was an enormous step forward for the development of science but this idea, although it originated with Sir Robert, had not come from Freemasonry. It was an idea which Moray conceived from his experience of sharing ideas with a correspondence circle. The germ of the idea must have started as Moray whiled away the long days of his imprisonment in Bavaria, writing to the German scholar, Kircher, about magnetism and Egyptian hieroglyphics. Moray was never much of a scientist but he had noticed that his grasp of the subject greatly improved when he wrote down his ideas and invited other interested people to comment on them. This method of improving his learning stuck with him and throughout his life he corresponded with scientists. His letters to Alexander Bruce give enormous insight into his Freemasonry and his letters to Huygens reveal his developing thoughts for the structure of his society. But by suggesting and encouraging the creation of *The Philosophical Transactions* Moray made another important contribution to the development of modern science.

The regular publication of the results of experiments speeded up the communication and the collation of scientific results in a way that Galileo could never have imagined possible.[30]

Historian R K Bluhm says of this venture:

> *The financial responsibility was his [Oldenburg's] from the beginning, and the fact that he probably started his journal as much for profit as from altruistic motives need not detract from the credit due to him.*[31]

Oldenburg published a new edition each month and when he had completed twenty-two editions he had them bound into a volume, to which he added a title-page and index. He dedicated this bound volume to the Royal Society. From that time on he created a new volume, of twelve editions, each year. He was editing the twelfth volume when he died, and it still sits as he left it, without a title page or index, in the library of the Royal Society. Once more Sir Robert Moray's skill in persuading other people to carry out his plans, at their own expense, can only be admired! By allowing Oldenburg this publishing concession the Society was able to retain his services as a full time Secretary, without actually paying him for his work.

After Oldenburg's death successive secretaries continued to

publish the journal each month as their own private money-making sideline. It was not until 1750 that the Society itself took on formal responsibility for the *Transactions*. But editing and selling the *Transactions* did not turn out to be a licence to make money. Oldenburg may well have had great financial hopes for his journal but he was unlucky in his timing. Soon after he started his regular publication the book trade in London was greatly depressed. The reason for this was simple. A lot of customers for books developed headaches, nausea, vomiting, aching joints, and a general feeling of ill health; then they died . . . of the plague. Most of the survivors moved out of the city and away from the booksellers. Now this response had nothing to do with the publication of the *Transactions*, but a lot to do with the public hygiene of the city. It is perhaps more of a wonder that Oldenburg continued to publish his transactions regularly during this dreadful period rather than a surprise that he did not make much money from selling his pamphlets. He did, however, continue to sell his literary and political intelligence to earn enough money to eat.

Conclusion

The blood-curdling oath of secrecy which was a part of Freemasonry in the seventeenth century may well have contributed to the lack of written evidence of Masonic activity recorded by the early founders. However, later writers had named early Freemasons and many early members of the Royal Society were on this list.

John Evelyn's writings and his dealings with William Oughtred suggested that Evelyn might also have been a Freemason. The story of Sir Robert Moray was now coming together like a jigsaw with only a few pieces still left to fit. The Masonic links of the Gresham Twelve had become clear and their list of early appointees also seemed to contain a great number of Freemasons.

The Society soon broke away from its Masonic roots and its non-Masonic members increased and began to carry out their own agenda. Moray's plan was a success in its early years, as the Royal Society was almost a technical department for the Navy Board. After the rejection of the First Charter Sir Robert Moray devoted his energy to encouraging means of speeding up scientific research and he thought up the very first academic journal, *The Philosophical Transactions*. Typically he managed to persuade somebody else to put up the time and money to run it.

CHAPTER 11

Gossips, Spies and French Mistresses

Our fleet is Divided; Prince Rupert being gone with about 30 ships to the Westward; as it is conceived, to meet the French, to hinder their coming to join with the Dutch. My Lord Duke of Albemarle [General Monck as was] lies in the Downs with the rest, and intends presently to sail to the Gunfleete.[1]

Samuel Pepys, June 1666

IN 1666 THE NAVAL BATTLE between the Dutch and English fleets was not going well. The French were now supporting the Dutch and on 1 June 1666 they managed to split the English fleet. Pepys's diary entry about the forthcoming battle opened this chapter.

Monck lost six thousand men and twice as many wounded in the battle that followed. Of his sixty ships he lost eight sunk and nine captured. It did not look good for England and failure to defeat the hated Dutch was undermining Charles's reputation as a king and leader.

The English fleet regrouped and, under Admiral Thomas Clifford, it sailed out to do battle with the Dutch off the French coast. By the end of July the Dutch and French had 160 ships set on fire and were forced to withdraw.[2] The military campaign had ground to a stalemate and talks about peace terms began.

Then at three in the morning of Sunday 2 September Samuel Pepys was awakened by Jane, his maid. She told him that she had seen a great fire in the City. Pepys got up and, slipping a nightgown over his nakedness, went to the top of the house to look out of her window at the blaze.

This was the beginning of the Great Fire of London and during that night 'the wind was great behind the fire' and it continued to spread. At 4 o'clock the following morning, Pepys was again woken by Jane and once more donned his nightgown, but this time to do more than sightsee. The fire was getting closer to his house and he loaded his money, his plate and his best things onto a cart, and still wearing nothing more than his nightgown, he set off with his household to the home of a friend, away from the path of the fire.

The king became personally involved in fighting the fire. He led his own guards out to pull down houses to create firebreaks to try to stop the fire spreading through the close-packed timber buildings of the city. Charles carried a pouch of gold to pay householders for the houses 'so they might be demolished for the common good'. But the fire, driven by the freshening wind, jumped the gaps and continued to rage.

It burned for almost five days and destroyed an area half a mile wide and one and a half miles long. Three quarters of the square mile in the centre of London was completely destroyed and its inhabitants made homeless.

Naturally scapegoats were sought. The Dutch and the Papists got equal shares of the blame. But there was an unexpected consequence. Henry Oldenburg was neither Dutch nor Papist, but he was a foreigner and a foreigner with access to state intelligence! Even during the war he had continued to sell items of interesting correspondence to third parties, as his letter to Boyle shows. As a potential scapegoat he had many useful attributes.

Two of the main customers for his foreign intelligence were

Lord Arlington and Sir Joseph Williamson. Historian Dr David Mackie says of this relationship:

> *Arlington, the Secretary of State, and Sir Joseph Williamson, the Under-secretary, were well aware of the nature of Oldenburg's extensive foreign correspondence, which was not as exclusively scientific as might have been imagined, and fully appreciated its value as a source of foreign news.*[3]

Williamson had been made a Fellow of the Royal Society on 5 February 1663 and he also served on the Council in 1666. The pressures of office, brought on by the Dutch War, forced him to leave the Council and he did not serve on it again until 1674.[4] But as a council member he had proposed a mutually useful arrangement between his Department of State and the Society. Michael Hunter described the relationship so:

> *Sir Joseph Williamson was able to assist with postage on foreign letters by placing diplomatic channels at the Society's disposal.*[5]

The system worked like this. All the overseas letters for the Society were addressed to Henry Oldenburg care of Williamson's office, so that the government paid the postage. Oldenburg would go to the Office of State (the equivalent of today's Foreign Office), and collect the letters. He would copy them out and pass back to Williamson any civil or political news.

Previous historians of the Royal Society have been puzzled as to why Oldenburg was arrested in June 1667. R K Bluhm said of the matter:

> *It is evident that neither the Secretary of State nor his Under-Secretary had any reason to wish Oldenburg in jail, because their interests were clearly better served by leaving him at*

> *liberty to continue his correspondence unhindered. It is highly probable that the arrest was on the direct orders of the King himself, the reason being in some correspondence that cannot now be traced.*[6]

To understand why the king should act in this way it is necessary to realise just what else was happening in England in the summer of 1667. The plague, the war with the Dutch and the Great Fire had all cost the king enormous amounts of money. Before the start of the war his annual deficit had been running at about £400,000 at year and by the summer of 1667 the king's general debts were estimated by the Treasury to be around £2,500,000, an enormous sum in those days. Over £1,000,000 of this was owed to the Navy for costs incurred in fighting the Dutch and the French. To add to the problem rumours were being spread, accusing the king of misusing the money Parliament had voted for the Dutch War. For the early part of 1667 Charles had been negotiating with Johann De Witt to try to end the stalemate into which the war had degenerated. Then, in June, De Witt's Navy carried out a raid that was to tip the balance in favour of the Dutch, greatly embarrass Charles and wrong foot him in the delicate negotiations.

Charles's reputation had been going steadily downhill during the Plague and the reversals of the Dutch War, but had been greatly restored by his very public actions during the Great Fire.

His enormous enthusiasm and support for building a new London had also endeared him to the refugees of the Great Fire. William Preston attributes this rise in the king's reputation to his Freemasonic activities and the support of prominent members of the Craft saying:

> *After so sudden and extensive a calamity [the Great Fire], it became necessary to adopt some regulations to guard against*

any such catastrophe in future. It was therefore determined, that in all the new buildings to be erected, stone and brick should be substituted in the room of timber. The King and the Grand Master, [Thomas Savage, Earl Rivers] immediately ordered Deputy Grand Master Wren to draw up the plan of a new city with broad and regular streets. Bro Wren was appointed surveyor general and principal architect for rebuilding the city, the cathedral of St Paul, and all the parochial churches enacted by parliament, in lieu of those that were destroyed. This Brother, conceiving the charge too important for a single person, selected Bro Robert Hooke, Professor of geometry in Gresham College, to assist him; who was immediately employed in measuring, adjusting and setting out the ground of the private streets to the several proprietors . . . On the 23rd of October 1667, the king, in person, levelled the foundation stone of the new Royal Exchange Building in due form . . . In the centre of the square, [within the new Royal Exchange] is erected the king's statue to the life, in a Caesarean habit of white marble, executed in a masterly manner by Bro Gibbons, then Grand Warden of our Society [Freemasonry].[7]

Three of the men Preston claims as senior members of the Craft are prominent members of the Royal Society. Charles II (Fellow 180), Christopher Wren (Fellow 12) and Robert Hooke (Fellow 136).

Returning to that fateful summer of 1667 Charles was about to be held up to ridicule by the Dutch. De Witt instructed his sailors to attack 'London's River', the Thames. With the help of two traitorous English pilots his fleet captured the fort of Sheerness and destroyed the boom at Chatham dockyards.

John Evelyn tells the story in his diary starting on 11 June 1667:

To Lond: alarm'd by the Dutch, who were falln on our Fleete,

> *at Chatham by a most audacious enterprise entering the very river, with part of their Fleete, doing us not onely disgrace, but incredible mischiefe in burning several of our best Men of Warr, lying at Anker & Moored there, & all this thro the unaccountable negligence of our not setting out our fleete in due time: This alarme caused me (fearing the Enemie might adventure up the Thames even to Lond, which with ease they might have don & fired all the Vessels in the river too) to send away my best goods and plate etc from my house to another place; for this alarme was so greate, as put both county and Citty in to a panique (panic) feare & consernation, such as I hope I shall never see more.*

The Dutch fleet blockaded the mouth of the Thames for two weeks until finally driven off at the Battle of Gravesend.

Evelyn, who was a strong supporter of the king was distressed by the Dutch blockade and clearly saw what damage it would do to the king's reputation. On 18 June he wrote:

> *I went to Chatham, and thence to view not onely what Mischiefe the Dutch had don, but how truimphantly their whole Fleete, lay within the very mouth of the Thames, all from North-foreland, Mergate, even to the Buoy of the Nore, a Dreadful Spectacle as ever any English man saw, & a dishonour never to be wiped off.*

But what has all this got to do with Henry Oldenburg? It appears that his private arrangements for passing on useful items of interest to Williamson had gone slightly awry. Lord Arlington had a very useful spy acting in De Witt's court, the playwright Aphra Behn. She had written a warning to the government about this raid, sending it via Williamson's office, to alert the king of the threat to the Thames. In her memoirs she wrote that her news 'might have sav'd the nation a great deal of money and disgrace had credit been given to it'.[8]

Why was her information ignored? The sudden arrest and unexpected imprisonment of everybody's favourite spy and gossip, Henry Oldenburg, during this blockade provides a possible clue.

Just before the arrest Evelyn had been discussing with Arlington why the king had been so badly advised: 'Those who advised his Majestie to prepare no fleete this Spring, deserv'd I know not what. I had much discourse with him, I told him I wondered why the king did not fortifie Sheerness.' Evelyn, who also purchased intelligence from Oldenburg, seems to have been aware of the threat and was surprised no precautions had been taken. How could he have known this when Arlington did not?

Evelyn's diary entry of 8 August adds to the story:

> *Home, by the way visiting Mr Oldenburg now close Prisoner in the Tower, for having been suspected to write Intelligence Etc: I had an order from my L. Arlington, Secr of State which made me be admitted: this Gent; was Secretary to our Society & will prove an innocent person I am confident.*

Was Mrs Behn's letter addressed via Oldenburg to avoid suspicion? Did Oldenburg fail to alert Williamson and Arlington to her message? Or did they overlook its significance and use Oldenburg's role, as general message handler, as a convenient cover for their own lack of care? Surely the fact that Arlington himself was suspected of being a Papist[9] would not have encouraged him to imprison an honest foreigner simply to direct attention away from himself?

Oldenburg admitted fault. He wrote to the king apologising for his 'neglect, which having given offence, I am ready to beg his Majesty's pardon for, upon my knees'.[10]

Whatever the motives for his imprisonment, the end result was that Oldenburg stayed in the Tower for ten weeks and was not released until the end of August, just after peace was negotiated with the Dutch.

This peace settlement laid the basis for the English success in colonising North America. Historian Lord Elton explains it:

> *By the treaty of Breda (1667) we surrendered Surinam in Guiana to the Dutch, but, what was vastly more important, the Dutch retired finally from the North American mainland. For the Dutch settlements which were to become New York and New Jersey had been captured in the course of the war, and were not returned. The Dutch Empire was destined henceforth to be an Empire of trading stations in the Tropics. The peopling of vast temperate regions from the mother country, the spread of their own way of life across the new continents, all this and how much more, had they only known it, they were abandoning to the English. An historic achievement to be ascribed to the government of Charles II.*[11]

The Royal Society had now fulfilled the role Sir Robert Moray had envisaged; it had built Charles a navy he could use to thwart the Dutch. But by now it was developing a life of its own and had more to contribute than Sir Robert had realised. Oldenburg went back to his correspondence and continued to publish the *Philosophical Transactions* each month. And the fellows gradually got into the habit of sharing knowledge by publication. This led to further useful spin-offs for Charles.

Navigating the World

North America was a long way off and regular contacts with the continent involved crossing and re-crossing the Atlantic, and this at a time when the determination of longitude was largely a matter of guesswork. Historian of science, Professor Herbert Butterfield, commented on this development:

> *Much of the attention of the Royal Society in its early years was*

> *actually directed to problems of practical utility. And for a remarkably long period one of the topics constantly presented to the technicians and scientists was a matter of urgent necessity – the question of finding a satisfactory way of measuring longitude... It has become a debatable question how far the direction of scientific interest was itself affected by technical needs or preoccupations in regard to shipbuilding and other industries: but the Royal Society followed Galileo in concerning itself with the important question of the mode of discovering longitude at sea.*[12]

In the developments of the science of navigation, and the subsequent understanding of astrophysics which followed there were three major players. These were Isaac Newton, Edmund Halley and John Flamsteed. But the story really starts with a woman, Louise de Kerouälle. She was an extremely attractive young lady in waiting to Henrietta Anne – Charles's sister. Looking at Henri Gascar's painting of her, long-haired, bare-breasted, stroking a dove with her long sensuous fingers as she gazes out of the picture with the enigmatic smile of a well-satisfied lady, it's easy to see why Charles made her his mistress. Charles fancied Louise from the moment he first saw her, during his sister's state visit in 1670. Henrietta prevented Charles from seducing Louise during that visit, but soon after Henrietta died. Louise returned to Charles's court still a virgin. Although to expect Charles to leave her long in this state would have been virgin' on the ridiculous. Charles nicknamed her Fubbs, naming his favourite yacht in her honour. His native harem was less respectful. Nell Gwyn first nicknamed her Squintabella and later 'the weeping willow' when she tried to bend the king to her will by the use of copious tears.

The king sometimes suffered from slight indispositions in his lady friends, which detracted from their ability to entertain him and he was not always happy about any resulting lack of availability. He accidentally gave 'poor Nelly' a dose of 'the pox', leaving her unfit for Royal service, at least until the inflammation died down.[13] To him it must have seemed a quite logical step to guard against any future inconvenience by increasing his stable of mistresses. A satirist of the time described the king's new brace of mistresses as Snappy and Tutty.[14] Snappy was Lousie, so called because she was given to scolding the king and Tutty was Nell, named after her low breeding, which caused the courtiers to go 'tut tut' at her. There is little doubt that Charles was susceptible to scolding. When Nell complained about him giving her a 'dose of the clap' he bought her a pearl necklace. Not to be outdone Snappy developed her sense of drama to the extent of threatening suicide if the king did not do exactly as she wished. Louise had quickly grasped that the king could easily be persuaded to do things if enough hysterics and tears were paraded before him. She understood Charles well. One of her ladies in waiting, Lady Cowper, wrote in her diary of how the king responded when being told that Louise would die if he did not go to her. Charles responded:

> *I don't believe a word of this; she's better than you or I are, and she wants something that makes her play pranks over this. She has served me so often so, that I quite sure of what I say as if I was part of her.*

Charles, however, was never able to keep up this stern demeanour should he be within arm's length of the lady. He always felt the need to console feminine tears with kind words, and probably firm deeds as well! Louise, became the Duchess of Portsmouth after successfully giving birth to a Royal Bastard (the Duke of Richmond). It does seem a rather thoughtless choice of title on Charles's part since he had married Queen Catherine in Portsmouth. But Fubbs frequently did want something when she played her pranks on the king and often what she wanted coincided with the interests of the French. Antonia Fraser hints that Louis XIV, King of France, intended to plant the virgin Louise on Charles for his own purposes:

> *The varied intrigues which led to the establishment of Louise [as Charles's mistress], the whole process of dangling this nubile beauty before the famously susceptible king, all presumed that Charles's political sympathies followed his amorous inclinations.*[15]

Historian Clive Aslet goes even further in his suspicions:

> *She was sent by Louis XIV, to promote what would later have been called an entente cordiale between England and France. It was her mission to meddle in politics and she did – not only in politics, but in any matter where her countryman's interests were affected.*[16]

It was an interesting side effect of Charles's lusty interest in Louise that resulted in the Prime Meridian going through Greenwich. That the most important vertical line on the maps of the world should be a monument to Charles's seduction of the Celtic beauty the French called 'la Belle Bretonne' seems somehow rather fitting.

'The Fair Lady Whore', as John Evelyn described Louise, persuaded the king to give an audience to a Frenchman by the name of Le Sieur de St Pierre. He told the king that he had discovered a way of determining longitude at sea by a simple application of astronomy. The king didn't understand the technical explanation he heard but was impressed enough to refer the matter to the Royal Society to investigate. A committee was formed under Sir Jonas Moore, who invited a young astronomer along to the meetings to contribute technical advice. That young man was John Flamsteed.[17] The committee met on 12 February 1675. It was Flamsteed who pointed out the two faults with St Pierre's method. Firstly, it relied on knowing accurate positions for all the fixed stars in the heavens, and secondly it didn't work. The Society reported back to the king, with a recommendation that as a first step towards any solution to the problem of longitude a detailed map of the heavens and the movements of the Moon would first have to be created. They suggested the king appoint his own Astronomer Royal, whose job would be to create such a catalogue.

The Royal Observatory

The king responded by ordering an observatory to be built and consulted the Society on who should run it and where it should be. Sir Jonas Moore suggested John Flamsteed for the appointment and Flamsteed was duly taken on. But where was the new Royal Observatory to be set up? Sir Jonas wanted to build it in Hyde Park, Flamsteed himself suggested the old Chelsea Hospital, which the Royal Society owned at that time, but it was left to Christopher Wren, himself no mean astronomer, to point out that the rural site of Greenwich would be well away from the fogs and smogs of London town. The king already owned the site, it was known as Greenwich Castle, and it had been rebuilt into a palace for his mother. It seemed an ideal spot, and the king agreed to Wren's suggestion, provided the works did not cost more that £500. Charles wanted a catalogue of stars but he was getting used to the idea that scientists would pay for the privilege of being allowed to practise their science in support of the king. Flamsteed, however, was not an independently wealthy man. Wren built an impressive eight-sided building which focused on the high-windowed observing room, known today as the Octagon room. Contemporary engravings show a building which looks rather like a castle from the outside.

Inside the Octagon room Flamsteed and his assistants made measurements through the windows using long thin telescopes. Looking at the etching suggests a well-staffed, well-equipped, state-of-the-art observatory, but unfortunately this was not the case. King Charles provided the building, and even paid Flamsteed £100 a year, but did not provide any money for equipment. The clocks, shown in the engraving were the gift of Sir Jonas Moore, while all the telescopes belonged to the Royal Society. Once more Sir Robert Moray's great scientific support organisation solved a public funding problem. Charles couldn't afford to equip the new Royal Observatory at Greenwich, so the more wealthy fellows came to his rescue by funding equipment, while the poorer scientific journeymen provided the labour and expertise. There was, however, one great drawback to this prestigious observatory as it is depicted, in 1676, in Francis Place's print entitled *Prospecus Intra Cameram Stellatam*. To save money Wren had used the foundations of the old Duke Humphrey's Tower, which meant the observing windows did not align with the observational meridian. Flamsteed solved the problem by doing most of his observations from a shed in the garden.

Small wonder Flamsteed felt so hard done by that he quarrelled with the Royal Society over the publication of his results. He quarrelled with Sir Isaac Newton and with Sir Edmund Halley to such an extent that his full star catalogue, more extensive than any which had previously existed, was not published until after his death on New Year's Eve 1719.

From Sir Robert Moray's first concept there were two classes of Fellow: those who were skilled in the sciences, able to conduct and theorise about experiments; and those who were not only interested in science but also had the money and social prestige to act as patrons of the working scientists. This policy worked during Sir Robert's lifetime but it was not to prove a firm foundation for the final success of the Society.

Conclusion

The overall picture of the early days of the Royal Society was now becoming much clearer. I was beginning to feel that I was starting to understand the unlikely events of the Society's formation and just how modern science had suddenly begun to flourish so soon after the strife and confusion of the Civil War.

The main clue to this new understanding was the character and motives of Sir Robert Moray. As I had studied him, and researched the intricate interlocking of the key events of his life, along with the tides of history which carried him along, a distinct pattern had started to emerge. I was becoming certain that Sir Robert Moray had been the main instigator of the Royal Society.

An interplay between the aims of one of the king's French mistresses and the Society's interest and expertise in navigation led directly to the formation of the Royal Observatory at Greenwich. This was an undertaking which the Royal Society supervised and helped to equip.

There really only remained one further question to ask. If Freemasonry and its teachings played such an important part in the formation of the Royal Society, why is the fact not more widely known? Today there are no links between Freemasonry and the Royal Society and while the Royal Society has developed as an important force in the modern world, Freemasonry, particularly in England, has atrophied and largely lost its way.

If I was correct about the Masonic origins of the Royal Society then I would have to explain how this knowledge was lost. When did the split occur and what caused it? This was the final matter I would need to investigate. To address it I would need to look more closely at what was happening to Freemasonry and the Royal Society during the events which followed the fall of the Stuart kings.

CHAPTER 12

A Legend of Gracious and Kindly Kings

On St Andrew's Day [1830], the Council of the Society was elected first, with mixed results . . . The voting was close, 119 for [the Duke of] Sussex, 111 for [scientist John] Herschel . . . So Sussex had won [the election for the Presidency of the Royal Society], in a contested election he had hoped to avoid.[1] Marie Boas Hall

THE STUARTS HAD COME from Scotland and made Freemasonry fashionable in seventeenth-century England. When the German Hanoverian line took over the throne of England they were not popular with the whole populace and the supporters of the exiled Stuart line became known as Jacobites. The problem that the Hanoverians had with Freemasonry was that it had grown from a Jacobite organisation, and they saw its Jacobite heritage and traditions as a threat to the stability of their own line. This simple political fact underlies the suppression of Freemasonry's Scottish roots. In 1717 it would have been dangerous to stand up in London and say, 'I am a Freemason.' Such a statement would be treated as expressing support for the recently defeated Jacobites and was a clear invitation to be hanged, drawn and quartered as a traitor to the Hanoverian crown.

Charles II died suddenly. His rapid descent into incapacity and death has often been attributed to a stroke but medical researchers Myron Wolbarsht and Daniel Sax put forward another suggestion.[2] They drew attention to the fact that Charles took an increasing interest in the process of fixing mercury as he got older. In his youth Charles was famed for his lusty ability in the bed-chamber. De Beer says of him:

> *His love of women was earthy; his seventeenth mistress abroad (if she was the last in the series) was succeeded after his return to England by a troop, sometimes, two or three at a time.*[3]

Once Charles was over fifty he began to have trouble satisfying even a single lady, let alone three at once. Wolbarsht and Sax link mercury vapour and its effects on blood pressure to his obsession with personally experimenting with mercury inhalation. In the short term the effects of sniffing mercury vapour would have acted like an early form of Viagra. It would have increased his blood pressure and with it the strength and durability of the Royal erection. In the slightly longer term it would have made his judgement totally unreliable and it would eventually kill him. The symptoms of his last illness are more typical of mercury poisoning than they are of a stroke but the increase in blood pressure caused by excessive exposure to mercury could, in itself have brought about a stroke.

If Wolbarsht and Sax are correct about the cause of his symptoms, and Charles was deliberately inhaling huge wafts of mercury vapour to try to prop up his failing love life, then his final deathbed conversion to Roman Catholicism becomes easier to understand. Throughout his life Charles had been careful to maintain an even-handed attitude to all religions and he had been very clear that he owed his restoration to his practice of the Protestant faith.

The symptoms of mercury poisoning are hallucination and

dementia. In 1685 the poisonous effects of mercury were unknown; indeed, it was widely used by hatters as part of the process of shaping felt. Mercury poisoning was an occupational disease for hatters and gave rise to the term 'as mad as a hatter'. It is well documented[4] that Charles took a particular interest in the properties of mercury during the last few months of his life. He had a well-equipped laboratory within the Palace of Whitehall, which had been originally set up by Sir Robert Moray, and he spent increasingly longer periods of time in this room towards the end of his life. Exposure to mercury vapour would have made him become quite demented and very susceptible to any suggestions made to him, particularly by those who were close to him, such as his brother James.

Religion had never played much of a role in Charles's plans, except as a necessary evil when dealing with fanatics. He had been flexible enough to take the Covenant of the Presbyterians in return for the throne of Scotland. In addition he was happy to return as Head of the Anglican Church, in exchange for the throne of England, Wales and Ireland. Despite this Charles was still able to convince Louis XIV that he would convert to Catholicism in return for French gold, but Louis never seemed to be able to provide quite enough gold for the spiritual alchemy to occur. It seems totally out of character that Charles should abandon his lifetime's studied indifference to religion in his final hours.

James, however, had a very different agenda from that of his elder brother. He welcomed a deathbed conversion for Charles, in the interests of restoring the Catholic faith to Britain. James made no secret of the fact he was a Roman Catholic. He had been forced out of his position as Lord High Admiral because he would not renounce his Catholic faith. James had visions of re-establishing Roman Catholicism as the official religion of Britain, something which had never been part of Charles's intentions. The deathbed conversion to Catholicism, which

James stage-managed ostensibly at the request of his hallucinating brother, seems to have been an opportunist response to the sudden susceptibility of the dying king.

James II

Sir Winston Churchill says of James:

> *James was a convert to Rome. He was a bigot, and there was no sacrifice he would not make for his faith. He lost his throne in consequence, and his son carried on after him the conscientious warfare, to his own exclusion . . . Protestant opinion has never doubted that if he had gained despotic power he would have used it for his religion in the same ruthless manner as Louis XIV . . . The English Protestant nation would have been very foolish to trust themselves to the merciful tolerances of James VII(II) once he had obtained the absolute power he sought.*[5]

James never achieved that absolute power, as he was driven out of Britain in 1688. During this 'Glorious Revolution' Parliament invited the Protestant rulers of Orange, William and Mary, to become joint monarchs. James had so little support in England that he was forced to flee to France with his wife and son, Prince James Francis Edward Stuart.

The honeymoon of William and Mary with Scotland would soon come to a violent end. Although Scotland had accepted the rule of William and Mary, not all Scots had abandoned James VII(II). Viscount Dundee led a Highland uprising to restore him to the throne of Scotland. But after 'Bonnie Dundee' was killed at the Battle of Killiecrankie this attempt to reinstate James melted away.

But just as William, the new Protestant king, was becoming accepted, he made a tremendous error of judgement. William issued a decree that all Scottish chiefs should take an oath of loyalty to him. One chief, MacIan of the clan Macdonald was three days late in taking the oath. William had issued a written order that anybody who failed to take the oath by the set date was to be severely punished. Two months later MacIan and his entire Macdonald clan were murdered by the men of clan Campbell in the Massacre of Glencoe, carrying out the letter of William's written orders.

All Scotland was horror-struck at the terrible crime. Men could scarce believe that a king could have given such an order. But it was true; although it is hard to believe that William intended that his law would be carried out in such a terrible way. The Scots never forgave the king for his carelessness and support for the exiled Stuart line was immensely strengthened. The Glencoe Massacre made many of the Scottish people begin to think they might be better off with a king of their own. William's standing with the Scots fell further when he failed to support Scottish attempts to establish new colonies in the Americas. All in all the Scottish people were no longer sure that they wished to be joined to England.

Meanwhile, in a last ditch attempt to regain his throne, James encouraged Catholic Ireland to rise to his aid and remove William, but he was defeated at the Battle of the Boyne and the Protestant Succession of 'King Billy' to the throne of England was confirmed. The supporters of James were driven underground and then took the name Jacobites, in honour of James. James himself returned to France, setting up a Court in Exile at St Germain, where he stayed until his death in 1701. His son, James Francis Edward, was recognised as James VIII(III) by Louis XIV. Freemasonry played a major role in supporting the Jacobite cause and as it is always the winners who write history, it was English Hanoverian Freemasonry which survived into the Masonic history books.

When William died the crown passed to Anne the daughter of James VI(I) and her heirs. The Scots Parliament then passed a law which said that the ruler of Scotland after Queen Anne had to be a different person to the one reigning in England. It was beginning to look like there would be war between Scotland and England and that the two crowns would be separated once more. In 1706, to try to avert this split, there was a move to combine the Parliaments of Scotland and England. This worried the Presbyterians. They had not forgotten Bishop Sharp and they were afraid that the Episcopalians would again try to persecute them.

When James VII(II) died, the existing Freemasonry was loyal to the Stuarts. James VI(I), Charles I and Charles II had all been Patrons of Freemasonry, and were all said to have become Freemasons. I have not found any evidence that James II was a Freemason, but many of his followers were.

As long as the Stuart line of Queen Mary II (wife of William of Orange) and Queen Anne continued, there was no real conflict of loyalty between Freemasonry and monarchy. The problems only began when Anne died in 1714.[6] Unfortunately she died without leaving an heir, and the next in line to the throne was her younger half-brother, the exiled Prince of Wales, James Francis Edward. If the Prince of Wales had been prepared to renounce his Catholicism then he would have been welcomed back to the throne of Britain, but he was not willing to do so. This James harboured the same dreams which had destroyed his father's reign. He wanted to convert England to the religion of Rome and this was something to which the English were strongly opposed. So the Protestant Parliament placed on the throne of Britain a German king, a man who spoke not a word of English!

Anne left a kingdom close to civil war. The Treaty of Utrecht had curbed the power of France, the Dutch were content with boundaries they could protect and the succession to the Spanish throne had been settled to the satisfaction of most of the participants. After twenty-five years of war none of the countries of Europe had achieved everything they sought but at least there was an uneasy peace. Britain was split between the Jacobites, who wanted to invite the Prince of Wales back, and those who would not accept a Catholic king at any price.

'Good Queen Anne' had broken the power of France to dominate Europe and had presided over a great expansion of British national strength, but as she lay on her deathbed it seemed for a while that she might declare the Catholic Prince of Wales her successor and plunge Britain back into turmoil. Only a last-minute intervention by the Dukes of Somerset and Argyll persuaded the terminally ill queen that she must declare her third cousin, the Elector of Hanover and the German son of Sophia, (daughter of James VI(I)), her successor. As her dying act she confirmed the Succession would remain Protestant.

The First Hanoverian King

George I was an unprepossessing king. Winston Churchill described him as an obstinate and humdrum German martinet with dull brains and coarse tastes.[7] He had little interest in Britain or in English politics, having only visited England once previously. He came to England simply because it was the only way to take control of the crown which luck had given to him. In 1714 it was by no means clear-cut that the German Hanoverian line would succeed in holding onto the British crown. There were many who hoped for a Jacobite Restoration. The German monarchs lacked the lustre and sparkle of the Stuarts. There was no pretence that the Electors of Hanover ruled Britain by Divine Right; they held office only because Parliament willed it so. The early years of the Hanoverians were to be a period that developed many of the modern techniques of Parliamentary Government for the first time. As the king could speak no English he did not chair the Cabinet meetings as Queen Anne had done. The chair of the Cabinet was taken over by the First Lord of Treasury, Robert Walpole, who came to play such an important role in government that he was known in a popular term of what amounted to abuse as the 'Prime Minister'.

When James II died in 1710, Louis XIV proclaimed Prince James Francis Edward Stuart to be James III of England. This was not acknowledged by the British Parliament and so in 1708 James the Pretender, with the aid of French troops, tried to take control of Scotland to make good his claim. He failed in this attempt and was driven back to exile in France. However, by 1715 there was such widespread dislike of the arrogant German king, particularly in Scotland where he was referred to as 'the Wee German Laddie' that it was estimated by Marshal Berwick that five out of six persons even in England were Jacobites.[8]

On 1 September 1715 Louis XIV of France died. James was now left without his long-time supporter and protector. If he was ever to take back what he considered his rightful crown, then he needed to act. Five days later the Earl of Mar raised the Jacobite flag over the traditional crowning place of the Kings of Scots in Perth. Within days he had an army of ten thousand men, waiting for the return of 'the king from over the water' and marching on London ready to welcome him back.

When James landed at Peterhead, on 22 December 1715, his supporters had already lost the battles of Sherriffmuir and Preston and had been forced back into Scotland. James arrived to join them bringing neither money nor ammunition. He was then taken ill and did not reach Scoon until 9 January 1716. It was his intention to be crowned King of Scots, as his great Uncle Charles had been in 1650. The ceremony, to make him James VIII of Scotland never took place, since the elders of the Kirk refused to crown a Catholic. Meanwhile the Duke of Argyll was fast closing in on Perth with his victorious Hanoverian troops. James and his Jacobite followers made a hasty retreat back to France.

The Suppression of Freemasonry's Scottish Origin

All English Freemasons are told that Freemasonry originated in London in 1717 when four lodges held meetings at:

> *The Goose and Gridiron, in St Paul's Churchyard,*
> *The Crown, in Parker Lane near Drury Lane,*
> *The Appletree Tavern, in Charles Street, Covent Garden,*
> *The Rummer and Grapes Tavern, in Channel Row, Westminster.*

It seems on a whim these four lodges just happened to decide to join together to create a Grand Lodge to rule the Craft of Masonry. This new Grand Lodge then went on to develop the worldwide fraternal organisation that Freemasonry is today. Anyone reading the Masonic Year Book of the United Grand Lodge of England can be forgiven for believing this to be true. After all, this official book contains a ten-page section entitled Outstanding Masonic Events. The first entry, in this list of approximately six hundred events, is the statement:

> *1717 Grand Lodge convened, Anthony Sayer Grand Master.*[9]

All English Freemasons are required to accept that what describes itself as 'the premier Masonic Institution' was founded by the inspired action of four gentlemen's dining clubs who had adopted the rituals of the stone masons' guilds for their own moral betterment. The picture of a group of noble gentlemen wandering around their local building-sites, asking the stone workers could they please join the labourers' trade union so they could learn from the Mason's rituals seems to have been taken directly out of a Monty Python script. And, in the light of what I have already explained about the background to Sir Robert Moray's experience of Freemasonry, it seems highly unlikely. But why should such an oddball story have ever arisen?

Scotland clearly had a noble Grand Master Mason from before 1602 when the Schaw Statutes affirm that the Masons of Scotland acknowledged Sir William St Clair of Roslin as their patron and protector.

However, the hereditary Grand Masters of Masons in Scotland had a very embarrassing history when viewed through the eyes of English Freemasons. The St Clairs had supported the crowning of Charles II of Scotland against the wishes of the Lord Protector. Roslin Castle had paid the price for this defiance when it was razed to the ground by General Monck. Scotland had continued to support the Stuart Line against the English and in 1715 the Scots had supported James VIII(III)'s attempt to regain his crown from the Hanoverian Line of English kings.

The Freemasons of London were worried. There was a climate of witch-hunting following the crushing of the Scottish army of James. The lodges of England had to have come from somewhere and the only bodies to issue warrants to form lodges prior to 1641 were the Scottish lodges, taking their authority from the Schaw Statutes of 1602. Anybody with Jacobite sympathies was suspect; the Freemasons had very clear links with Scots who in turn had demonstrated a strong animosity towards George I. The four London lodges which met at the Goose and Gridiron, the Crown, the Appletree Tavern and the Rummer and Grapes Tavern were probably acting according to warrants originally issued by one or other of the Scottish Schaw Lodges, the only legitimate source of Masonic authority at the time.

For Hanoverian supporters this must have been incredibly disturbing, they would have known that for many years prior to the 'Fifteen' campaign the Scottish lodges had kept a fund to which all candidates contributed to provide for the purchase of weapons 'keeped and reserved for the defence of the true Protestant religion, king and country and for the defence of the ancient cittie and their privileges therein' and they were obligated 'to adventure their lives and fortunes in defence of one and all'.[10]

If London Freemasons wanted to continue to meet they would have to ensure that they purged their movement of its dangerous Jacobite associations but they had the problem that their authority to act as Freemasons stemmed from the obviously Jacobite Schaw Lodges of Scotland. Their solution was novel and almost certainly Masonically illegitimate; but they needed an alternative source of authority for their activities. They created such an authority by bringing together four London lodges, denying their Scottish origin and forming a Grand Lodge to govern all Freemasonry, except themselves. They then set about courting the Hanoverian Royal Family, encouraging them to join and eventually lead London Freemasonry. Within four years they had a noble duke as their Grand Master, within sixty-five years they would have a surfeit of Hanoverian Princes at their head. The price to be paid for this acceptance was the erasure of all trace of their Scottish roots (which in effect meant protesting no knowledge of any Freemasonry prior to 1717, a line they still adopt).

Not all Masons agreed that London Masons should rule all Freemasonry because almost immediately after the formation of the Grand Lodge of London, Grand Lodges were formed in Munster and Dublin to protect the interests of their Brethren, who were largely Jacobite in attitude. In Scotland the traditional self-ruling and warrant issuing lodges did not see the need to act but Scottish dissatisfaction with the English Hanoverian monarchy was growing.

There were a large number of associations formed to promote the interests of the 'king over the water' and his heir. Among these was the Royal Company of Archers of Edinburgh. By 1724 its activities with parades, competitions and shows of strength were worrying the insecure government of George I.

The threat posed by the exiled James VIII was causing much concern and so when the names of the inner group known as the Sovereign Bodyguard of Scotland were published by an English sympathiser, the Masons of Scotland could not avoid noticing that their hereditary Grand Master Mason was a Brigadier of the Jacobite Royal Company of Archers.

The Lodges of Scotland became concerned about the response of the Hanoverian Pretenders to Freemasonry in England, and to the formation of a Grand Lodge in Ireland. To prevent Wales forming its own uncontrolled National Grand Lodge in the same way, the London Grand Lodge offered Hugh Warburton the Office of First Provincial Grand Master and he in return sold the nation of Wales to England as a Province (a strange arrangement which still upsets many brethren in Welsh lodges). A system of control and patronage was quickly being developed to ensure all lodges complied with the edicts of the gentlemen Freemasons of London. The appointment of Brother the Earl of Strathmore, soon followed by Brother the Lord Crawford as Grand Master of London Freemasonry, suggested it would not be long before a Scottish Freemason would be found to be the first Provincial Grand Master of Scotland, also as a province of England. The Lodges of Kilwinning and Scoon and Perth did not think this was a serious threat but the Edinburgh lodges took the threat seriously enough to come up with a solution. They proposed to elect their own Grand Lodge to administer their affairs, issue warrants and protect their interests. To carry out this plan they needed a Grand Master Mason in which matter the Schaw Statutes left them no choice.

William St Clair of Roslin was their hereditary Patron. Accordingly he was initiated into Freemasonry on 8 May 1736, on 2 June he was raised to the sublime degree of a Master Mason and on 30 December he was installed as the First Grand Master Mason of Scotland. His very first act was to renounce and resign in writing his hereditary rights of Patronage and institute the system of election of Officers of Grand Lodge that still protects the rights and privileges of Scottish Freemasons.[11] Even Gould, whose major work on Masonic history has been carefully arranged in its second edition to play down the Scottish influence on the early history of English Freemasonry, grudgingly comments:

> *... the opportune resignation of William St Clair was ... calculated to give the whole affair a sort of legality which was wanting in the institution of the Grand Lodge of England.*[12]

The battle for legitimacy was now well underway. The Freemasons of London wanted to become loyal Hanoverians while much of Scotland stayed quietly Jacobite, only toasting the king after passing their hand over the glass.

After the 1715 Rising the Whig Government branded Tories and Freemasons as Jacobites and disturbers of the peace. It is small wonder that the Freemasons of London made strong efforts to distance themselves from the Stuarts by forming their own Grand Lodge in 1717, under the unfortunate Anthony Sayer. I use the term unfortunate advisedly, because as soon as the reborn Hanoverian Freemasons succeeded in attracting Anglo-German noblemen into their senior ranks they dumped Sayer and he survived only by acting as a paid Tyler (the guard who stands at the door of the lodge with a drawn sword while a meeting is in progress) for his own lodge. Clearly, it was not a good thing to have held rank within Freemasonry prior to 1717.

After 1717, new lodges quickly sprang up on the Continent, some founded by the Hanoverian Masons and others by the refugee Jacobites, but as all lodges welcomed brother Masons, without regard to religion or politics, these lodges quickly became sources of intelligence for both sides. Unfortunately for James, now known as the Old Pretender, Walpole was far better at the spying game than the Jacobites and James came to regard the Freemason's lodges which followed his court, first at St Germain and later in Rome, as threats to his chances to regain the crown of Britain.

Papal Condemnation!

Historian Alex Mellor sees this leakage of intelligence as the main motive behind the first Papal Bull against Freemasonry, which was issued in 1737. He says:

> *Freemasonry was divided into two tendencies: the one favouring the Stuarts being mostly Catholic, and the other favouring the House of Hanover being wholly Protestant; from this came a duel of espionage in which the Stuarts were not fit to fight, since they had neither the resource of a Walpole nor the intelligence of a Chesterfield. The day came when the Pretender James III, known as Chevalier de St George, finding that he had lost the battle in this field, gave the Holy See to understand that Freemasonry was no longer to be treated kindly, and that the interests of English Catholicism, incarnate in him, required the Church to file through the chain that held the ball. It did so.*[13]

It did so by issuing the first Papal Bull condemning Freemasonry. However, much of the Stuart Scottish support came from the Freemasonic lodges of Scotland and with Craft Freemasonry now a no-go area for Jacobite Catholics some other way of harnessing this Masonic under-swell of support, which was secure from Hanoverians, had to be found. The Stuarts did not abandon the Craft, instead they created the Royal Order of Scotland. It was open to all Master Masons and was set up, so they said, 'to correct the errors which had appeared in St John's Masonry in recent times'.

The Masonic Royal Order of Scotland can be reliably traced back to 1730, when a Chapter met in Charing Cross,

London.[14] It is a distinctly Jacobite organisation which still insists that its Grand Master has been and will always be the King of Scots. If the current monarch is not a Mason (as is the case at present), then an empty chair is kept at its meetings until such time as it should please a future Monarch to take that chair.

The traditional history, which the Order vouchsafes to its initiates, says that its first degree was founded by David I of Scotland (1124–53) and was first worked at Icomkill and later at Kilwinning. After the Battle of Bannockburn King Robert I added to the order another degree which confers civil knighthood on its members. The Order celebrates the Divine Right of the Stuart line to rule Britain and its rituals remind members of the Royal Order of their loyalty to the Kings of Scots.

This purely Jacobite branch of Freemasonry was only open to those who acknowledged the King of Scots as their Grand Master and was thus secure from Hanoverian Masons. This was the political answer to the problem of James the Pretender: that he dare not openly disown Freemasonry since many of his supporters were Masons. Alex Mellor sums up his problem:

> *James III (VIII), however, could not have ordered the exclusion of Masonry without foregoing his royal welcome to all Englishmen. He would have disavowed the monarchic principle. He would also have made sworn enemies.*[15]

The key line in the papal encyclical *In eminenti* published by Clement XII on 4 May 1738 gives the clue. The Pope condemned Freemasonry for two reasons. The first because Freemasonry encouraged its members to keep secrets from the Church and the second 'for other just and reasonable motives known to Us'. This second reason was political. Walpole was using Freemasonry as a political and intelligence tool against the Roman Catholic Pretender to the throne of Britain. It is not difficult to speculate that the reasons were not specified because to do so would have undermined the position of James VIII, whose cause the Holy See was trying hard to promote.

The confirmation of this speculation can be seen in the actions of Bonnie Prince Charlie. In 1747, after being symbolically crowned with a Laurel wreath at Holyrood, during the 1745 Rising, he declared himself to be Sovereign Grand Master of the Order. Prince Charles Edward Stuart thus distanced himself from his father James, who had disowned Masonry and become distrusted by his followers. So Prince Charles Edward became the patron of the most romantic order of Scottish Freemasonry and, unfortunately for the Hanoverians, Scotland has always been the fairy land of Freemasonry and so it would continue as a focus of discontent against the Hanoverian kings.

By 1746, after his failure to invade England, Bonnie Prince Charlie was routed by the Hanoverians at the Battle of Culloden. While his supporters were slaughtered or scattered, Charles fled 'over the sea to Skye' aided by the famous Flora Macdonald and made his way back to France. He became the 'Young Pretender' when his father died in 1766 and he styled himself Charles III, until his death in Rome on 31 Jan 1788. He left behind a younger brother, Henry, who was a Cardinal of the Roman Catholic Church and a number of disputed heirs to the Jacobite lineage.

Winston Churchill said of the continuing Jacobite tradition:

> *The Stuart's were to linger in men's memories as a sentimental, though ill founded, legend of gracious and kindly kings.*[16]

But the Stuarts had also left behind a powerful Masonic tradition which the Hanoverians viewed as a continuing threat.

The Grand Lodge of the Antients

In London, however, the Hanoverian Masons had not had everything their own way in their attempts to distance themselves from their Jacobite roots. A Mason brought up in the Irish tradition moved there in 1748 and joined a London lodge. He was so appalled at the changes this self-appointed Grand Lodge was making to the Freemasonry he had learned in Ireland that he decided to do something about it. This outspoken Irishman was Lawrence Dermot. Dermot was not well liked by his opponents as this description of him, written by a Hanoverian Mason shows:

> *As a polemic he was sarcastic, bitter, uncompromising and not altogether sincere or veracious. But in intellectual attainments he was inferior to none of his adversaries and, in a philosophical appreciation of the character of the Masonic Institution, he was in advance of the spirit of his age.*[17]

Dermot was born in 1720 and was initiated into the Dublin Lodge No 26, on 14 January 1740. He became Right Worshipful Master of his lodge on 24 June 1746. This was the year after Bonnie Prince Charlie's terrible defeat at Culloden. Dermot came to England in 1748 and joined London Freemasonry. He was a well-read man who spoke Hebrew and Latin and he was also a keen student of Masonic history. He believed that the attempts of the London Masons to accommodate the Hanoverian monarchy was forcing Freemasonry to move away from its Antient roots. Rituals and philosophy were being sacrificed by the Order in its attempts to distance itself from its Jacobite origins.

On 5 February 1752 he met with a group of like-minded Masons at the Griffin Tavern, Holborn. Together they set up an alternative Grand Lodge to oppose the assumed authority of the Grand Lodge of London. Dermot was made Grand

Secretary. They dubbed the Hanoverian apologists, controlled by the four London lodges, the Moderns. There is no greater term of abuse within Freemasonry than to be called modern and innovative. Dermot was determined to preserve the older Scottish tradition of Freemasonry. The Grand Lodge he formed was known as the Antient or the Atholl Grand Lodge, after the Duke of Atholl, a Past Grand Master Mason of Scotland, who became Grand Master during the period of their greatest successes.

To the Hanoverians, however, Jacobite Freemasonry posed an ongoing difficulty. They had solved the problem of securing the loyalty of the Hanoverian branch of Freemasonry by becoming its Grand Masters, but the Antient lodges were still suspect. Jacobite lodges had spread to North America and after the American War of Independence rumours abounded that the Freemason President George Washington had asked Prince Charles Edward, Grand Master of the Masonic Royal Order of Scotland, to become Charles III, king of the Americas. Whether or not it was true, even the rumour was a great embarrassment to the government of George III, who was by this time starting to show signs of dementia. The rumoured meeting between Washington's envoys and Prince Charles Edward Stuart was said to have taken place in Via San Sebastiano, in Rome, during November 1782. The disastrous loss of the rich American colonies was blamed on mad George III and in England dissatisfaction with the Hanoverian line started to grow once more.

Within ten years, however, there was another event which was to make the Hanoverians become even more afraid of the influence of Jacobite Freemasonry. In France a great revolution took place and the Jacobite Masonic lodges all told the story of how a French Freemason climbed on to the guillotine to wave aloft the severed head of Louis XV, shouting aloud. 'Jacques de Molay, thou art avenged at last!' (de Molay was the last Grand

Master of the Knights Templar, whose Order has been preserved within Freemasonry as the Masonic Knights Templar or KTs.) Popular writers of the time, such as Le Franc and Abbe Barruel, wrote great tomes exposing the role of the Freemasons in overthrowing the tyranny of the French monarchy. There is one particularly famous story of how Freemason François Westerman had led 600 followers towards Paris, to depose the king, and as they marched they sang the Masonic Anthem known as the *Chant deguarre pour l'armee du Rhin* written by Freemason Rouget de Lisle. We know that song today as *La Marseillaise.*

In Ireland the Protestant Loyalist Freemasons took the structure of the Masonic Lodge and created the Quasi-Masonic Orange Order. They said that the purpose of the Orange Order was to protect their Protestant Faith, but their rituals and myths recalled the glories of the Stuart kings and the successful battles of William and Mary Stuart against the Catholic James II. George III could take little comfort in this loyal celebration of the Stuart line.

As Britain went to war with France in 1797, the country was also threatened by rebellion in Ireland. By this time France was led by Napoleon Bonaparte, most of whose immediate family were Freemasons. When rumours spread from Egypt that Napoleon had been initiated into Freemasonry in a special lodge convened for the purpose inside the Great Pyramid of Giza, it is small wonder that the British Government decided to act against secret societies which threatened the stability of the Hanoverian Monarchy.

Unlawful Societies

In 1799 'The Unlawful Societies Act' was brought in by Prime Minister William Pitt for the 'more effective suppression of Societies established for Seditious and Treasonable purposes'.[18] Initially only Hanoverian Freemasonry, the Moderns, was exempted from this Act, as the Prince of Wales was Grand Master Mason of England. However, even Hanoverian Freemasons were required to register their names with the authorities when they joined a recognised lodge. A final clause which exempted all lodges of Freemasons, both Antient and Modern, from the provision of the Act was eventually agreed but there was a price to be paid.

On 18 May 1813 Prince Edward, who was also a leading member of the Moderns, wrote to the Duke of Atholl saying:

> *On every occasion we would be happy to co-operate with you in exerting yourselves for the preservation of the Rights and Principles of the Craft and that, however desirable a Union might be with the other fraternity of Masons, it could only be desirable if accomplished on the basis of the Antient Institution and with the Maintenance of all the rights of the Antient Craft.*[19]

The Antients agreed the terms of a surrender and submitted themselves, once more, to the authority of the Hanoverians. On 8 November 1813 the Duke of Atholl resigned in favour of the Duke of Kent. The scene was set for the imposition of total control on English Freemasonry by a United Grand Lodge of England. The Duke of Sussex was chosen to enforce this subjection by patronage, which has since become the hallmark of English Freemasonry. Politically the duke was a good choice as he was a prince of the Hanoverian blood and had married into the oldest bloodline of Freemasons known in England, the Moray family. (In 1793 he married Lady Augusta Moray, Countess of Inverness. Although a later Royal Marriage Act was to declare this marriage void he seems to have continued to live with the lady.)

At the time leading Masonic writers from other traditions, writing in the magazine *Freemason's Quarterly*, warned of the dangers of making him Grand Master:

> *Patronage it has been said, implies subjection, which latter, it is again urged can work no good to the Fraternity. Royal Brethren cannot but make their exalted position felt in the Lodge and thus affect the brotherly quality existing amongst members.*
>
> *Neither the English writer nor the English reader can keep clear from the egotistical insular tendency to look upon England as the central point of the whole system of events in this wide world.*

Gould's *History of Freemasonry* says of these events:

> *It has been truly said that the Duke of Sussex's whole heart was bent on accomplishing the desideratum of Masons, the Union of the Two Fraternities who had been mistermed Antient and Modern; his high station in life certainly carried with it an influence which could not have been found in an humbler individual . . . How far his plans were consistent with the original plan of the Masonic Institution, must be left to other judges to determine.*[20]

To avoid embarrassment to the Crown during the process of threatening the Antients with abolition by Act of Parliament, the Prince of Wales had appointed the Earl of Moira as Acting Grand Master. Gould again comments:

> *The Freemasons of England owe a deep debt of gratitude to the Royal Family of their country. Their immunity from the Secret [Unlawful] Societies Act of 1799 was due, in great measure, to the circumstance of the heir to the throne being at the head of the Grand Lodge of London (the Moderns). Later, when under the combined influence of two Princes of*

> *the Blood, discrepant opinions had been made to blend into harmonious compromise.*[21]

The Prince of Wales and his younger brother the Duke of Kent had forced the Antients back under the Modern's control but they left the task of rewriting the history of Masonic origins to the Duke of Sussex. To make sure he was obeyed he had every master of every lodge swear an oath to obey all edicts of his new United Grand Lodge of England (UGLE) before allowing that Master elect to take the Chair of his lodge. This regulation is still active today, being strictly enforced at every Installation of a Worshipful Master of an English lodge.

Sussex also realised that a weakness of verbal tradition is that it can be changed if the changes are made gradually. Knowledge of the original ritual and history will fade away. But patronage was his main weapon; to this day all officers of UGLE are appointed by the Grand Master, not elected by the Members.

The Duke of Sussex decided all things Scottish were to be discredited and the comment written by Masonic historian W A Lawrie in his *History of Freemasonry* in 1859 is typical:

> *It is certain that Scotland has always been the fairyland of foreign Freemasonry. Scottish was a good term to apply to any new degree.*

And yet in the proceedings of the Masonic body known as the Supreme Council of England in April 1909 a set of degrees which had been called the Scottish Rite was changed to the Antient and Accepted Rite. The minutes record:

> *it was decided to omit the word Scottish from all certificates etc.*

A reporter for *The Masonic Newsletter* at the time commented that this was a foolish resolution to pass as The Rite had always been called the Scottish Rite and was known as such in every corner of the world. To have set up a system which could make such a change shows that the Duke of Sussex had a complete disregard for Masonic history. He intended to totally suppress any possible Jacobite sympathies and by removing a sense of Stuart history from the Order he turned it into a political support mechanism for the Hanoverian monarchy.

The Duke of Sussex was the younger brother of King George IV and the son of the Mad King George III. In 1813 he had been created Grand Master of the newly formed United Grand Lodge of England, against the wishes of at least half the Freemasons in England. He was to retain the position until 1843. Then in 1830 he was narrowly elected President of the Royal Society as a result of a campaign on his behalf by the 'Gentlemen Patrons' of the Society and largely against the wishes of the scientists, represented by John Herschel. These two facts were to be my key to understanding why the Masonic roots of the Royal Society have been forgotten. Between 1830 and 1838 the Duke of Sussex was in autocratic control of the only two Societies which could have retained any record of the Jacobite Masonic roots of the Royal Society. I knew that the Duke of Sussex had misused his position in Freemasonry to suppress all Scottish, and hence Jacobite references, in its rituals and records. Had he also done something similar to the Royal Society?

Freemasonry's Missing Histories

There are two documents mentioned in Masonic verbal tradition, and recorded by Preston, which if they had existed within the Royal Society's papers could have been extremely embarrassing for the Hanoverian Duke of Sussex. They were both attempts at definitive histories of Freemasonry, written by two of the founding members of the Royal Society, our old friends, Sir Robert Moray and Elias Ashmole.

Historian D C Martin says of Moray's *History of Masonry*:

> *When selected Fellows were each asked to write up the history of a trade Moray undertook to prepare that on Masonry. This seems to have been done but is not now in the Society's records.*[22]

Moray was an expert on the symbolism and philosophy of Freemasonry, as his extensive letters to Brother Mason, Alexander Bruce show. He was also fully aware of the Scottish origins of the Craft and, I believe, aware of Charles II's membership of the Craft. Any history of Masonry this man wrote would be well worth reading but may well have contained references and disclosures that would have been unacceptable to a Hanoverian Grand Master Mason, set on denying the Jacobite roots of what had become a loyal Hanoverian Order. We can only imagine what Moray may have written in this lost work but Stevenson helps us appreciate the loss:

> *Sir Robert Moray cannot be taken to be a typical mid-seventeenth-century Freemason: the fact he reveals so much about what Masonry meant to him in itself makes him unique.*[23]

Ashmole decided to write his *History of Freemasonry* after the death of Sir Robert Moray. Ashmole had already written a complete history of the Order of the Garter and he is known to have had a completed manuscript of his *History of Freemasonry* before 1687. In 1682 Ashmole was listed in the records of the Royal Society as donating to the Society a 'number of treatises'. Was a copy of his treatise on Freemasonry among these papers? If so, it is no longer to be found at the Royal Society's library. Josten, writing in 1966, says in 1747 Dr John Campbell mentioned in a biographical article about Ashmole, that his manuscript on Freemasonry existed. Josten then goes on to comment that unfortunately that manuscript cannot now be traced.[24]

So two potentially embarrassing *Histories of Masonry*, either of which could have confirmed Masonry's Scottish Jacobite roots and its influence on the formation of the Royal Society have disappeared. I knew that the Duke of Sussex had shown he was totally capable of destroying Masonic material which contained Jacobite material. Did he have any opportunity to dispose of either Moray's or Ashmole's missing Masonic histories? To answer this question I needed to look more closely at the circumstances of the Duke of Sussex's election to the Presidency of the Royal Society.

A Royal President for a Royal Society

When Sir Robert Moray had conceived the original idea of the Royal Society he had envisaged two classes of members. The gentlemen patrons, who had money and were interested in science, and the skilled practitioners of science, who did not necessarily have wealth or in some cases, even a job. He brought together amateur patron and working scientist in a team which worked very well in those days before state funding for scientists.

In 1663 thirty-two per cent of the fellows were scientists, the rest amateur patrons. This proportion hardly varied from the granting of the first charters until 1830. On average, for its first two hundred years, the Royal Society seemed to need two and a half wealthy amateurs to support one working scientist.

However, at the beginning of the nineteenth century there was a dramatic change in attitude. It represented a move away from a belief that scientific research was a private matter, to be financed by patronage, towards a tendency for a more professional approach to scientific employment. Although the proportion of scientists had remained the same the total number of fellows had increased (in 1830 there were about 200 scientists and over 600 amateurs). This new class of professional scientists saw themselves as hindered by the amateurs, and by medical quacks, who they thought were swamping the scientists of the Royal Society. In particular the rules of the meetings where scientific papers were read did not allow any discussion of the findings. They were little more than a chance for the amateurs to meet socially and then dine together afterwards. To get away from this stilted dining club atmosphere many purely scientific societies were being formed such as the British Association for the Advancement of Science, the Geological Society and the Astronomical Society. Leading Scientific Fellows of the Royal Society were worried by what they saw as a decline in standards. William Herschel, the astronomer, said of British science at this time:

> *Our day is fast going by, and as we are both proud and poor and negligent we are rapidly dropping behind.*[25]

Thomas Young did not agree that all was lost and wrote in answer to him:

> *I fully agree with you that we are poor and proud as a country and too negligent of other nations; but I hope you are mistaken that our day is fast going by: for I do not comprehend that our scientific reputation has ever depended on the caprice of a ministry or its agents. What had King James to do with Newton's Principia?*[26]

Historian Marie Boas Hall says of this period:

> *The Cassandras announcing the decline of science in England thought of themselves as advocates of the professionalisation of science, and to a certain extent they advanced this cause, but it*

was chiefly to advance by means very different from those they advocated.[27]

One member of the Royal Society was a supporter of the Duke of Sussex's ambitions to be President, this was the current President at the time, Davies Gilbert. He was opposed to any attempt to allow the scientists control, and had said that he hoped that 'some honourable and noble personage' would come forward to offer himself as President. Using Moray's ancient rule that persons above the rank of Baron could be admitted on the day of application, Gilbert had sneaked the Duke of Sussex into an instant Fellowship with a view to making him the 'obvious' next President. Immediately after this move the Council amended the statutes to end this abuse of privilege and required a period of at least a month between proposal and voting. But it was too late, Gilbert had already got his preferred successor into the Society.

The Duke hoped that his exalted rank would ensure an easy succession. He wrote to the society, late in 1828, about the Presidency:

> *If proposed, I shall certainly accept it [the Presidency] and do my best . . . I should imagine they would wish to get me on the Council first and if so I should not object.*[28]

However, his internal mole, Davies Gilbert, consulted other members of the Council and when he found they did not want a Royal President he took fright and backed away from proposing Sussex as his successor. John Herschel was beginning to win more support among the scientists. Towards the end of 1828 he wrote:

> *What will the civilised world say to the cavalier kind of way which science and men of science are treated in England!*[29]

In April 1830, Charles Babbage (the inventor of the Calculating Engine which is the forerunner of the modern computer) published *Reflections on the Decline of Science in England.* In this book he attacked the amateurs who dominated the men of science in the Royal Society. Among a general torrent of abuse he also proposed that Fellowship of the Royal Society be restricted to practising scientists only. The Royal Society at first officially ignored his book but, when pressed by newspaper articles, eventually Gilbert issued a mild statement, regretting Mr Babbage's publication and disapproving of the 'uncandid' spirit in which it was written. An acrimonious correspondence developed in *The Times* and by the end of 1830 it was clear Gilbert would have to resign as President.

The Duke of Sussex's spin doctors started rumours of new royal patronage for the Society but it was not enough to save Gilbert. By September he was forced to resign. Secretary Pettigrew copied Gilbert's letter of resignation to Sussex and asked if he would stand as President. Sussex replied that he would be 'proud in filling the situation'. Gilbert now believed he had a *fait accompli* but he was surprised when the Council opposed the idea that the President could choose his successor.

Maria Boas Hall says of Sussex's attitude at this time:

> *Sussex himself was in an ambiguous position; he was interested, he was even anxious to be President . . . but he was inevitably far too royal to accept opposition and he was dependent on Gilbert.*[30]

At the Council meeting of 11 November 1830, the secret negotiations between Gilbert and Sussex were brought out into the open and the Council passed a resolution saying that all officers should be elected from among members of the Society who by their acquaintance with the conditions and interests of science were best qualified. This Sussex was clearly not! A whole flurry of scurrilous pamphlets now appeared accusing everybody involved with the most heinous crimes against the purity of science. It seemed that no one would dare to stand against the Royal Duke, but at the last moment, just before the St Andrew's Day election, John Herschel did. At first he did not want to oppose the Duke but when sixty-three of his supporters took an advertisement in *The Times* urging him to do so he decided it was his duty.

The voting was close. Sussex won, but only by 119 votes to 111, with two-thirds of the fellows abstaining. It was not an overwhelming vote of confidence in the Duke of Sussex but he had achieved his aim. This, however, was to be the last success of the non-scientific majority over the scientists, because the scientists had by now achieved a majority on the Council.

Sussex was pleased with the outcome; now he controlled both of the Institutions which had potentially embarrassing Jacobite records. He acted quickly. At his first council meeting on 9 December 1830, the week after his election, the new President took an immediate hand in the affairs of the Society. As Lyons says:

> *At this first meeting of the new Council, the President introduced a resolution, which was adopted, to the effect that the report of the last audit and the Treasurer's accounts for the twelvemonth ending 30 Nov 1830 should be printed and distributed to the Fellows forthwith . . . It seems that this was not on the agenda of the meeting but that the President made the proposal on his own initiative . . . Proposed, as it was, from the Chair by the newly elected President, no one ventured to oppose the motion.*[31]

The reason for this pre-planned action is not immediately clear from the accounts, which Sussex insisted were circulated to the Fellows. The accounts show that the Society was not short of cash; however, at the time there was a ground swell of support from the scientist Fellows to improve the Society's library of scientific works and to establish it in a better equipped reading room. This, the Duke of Sussex, for reasons of his own, was to do.

By the following year the accounts show a surplus of £775–7s-9d and this despite spending nearly £1,500 on the new library and over £500 on new scientific books to put in it. The money was raised by selling off many of the Society's less important papers, including the Arundel manuscripts, which were sold to the British Museum. Sussex had won the support of the disgruntled scientific fellows by allowing the scientists to see the Society's accounts for the first time and then explaining that the sale of the older, 'less useful' manuscripts to the British Museum would provide funds to build a modern scientific library. He appealed to their vanity saying:

> *I believe the scientific character of this country to be most intimately associated with the scientific character and estimation of the Royal Society.*[32]

To this end Sussex is remembered as the President who presided over the very first listing of the collection of books and papers of the Royal Society, although he seems to have quarrelled with Anthony Panizzi, an assistant librarian at the British Museum, who produced that first printed catalogue about what it should contain. The President took a great interest in identifying non-scientific books which were in such a bad state of repair as to be useless. He estimated this class of book to be about a third of the collection. Sussex 'secured additional rooms to house the library' and agreed to the disposal of 'useless' material.[33] No scientific material was either sold or disposed of. If any histories of Freemasonry had been kept by the Royal Society, as non-scientific books, they would have been automatic candidates for disposal.

While he was carrying out this modernisation of the Society's library Sussex seemed to be encouraging more and more scientists into the higher ranks of the Society. Marie Boas Hall says of this period:

> *At the end of Sussex's Presidency the officers were all strictly scientific men of some repute.*[34]

The scene was now set for the final transformation of the Royal Society into the premier scientific institution it is today. But why had Sussex been so eager to become President if he wanted to preserve the status of the amateur Fellow? His actions show he was not opposed to improving the scientific skills of the Society, so if he had not wanted to preserve the status of the aristocratic dilettantes why had he worked so hard to become President? I couldn't help thinking that his plan to 'modernise' the library, a task to which he devoted considerable time and effort, had to figure largely in his plans. His first move as soon as he was elected was to force the issue of the sale of non-scientific manuscripts to the forefront of Fellow's attentions, by circulating full financial information for the first time. The outcome of his actions resulted in the move to a wholly scientific society. For this he should be recognised but I could not help suspecting that this was yet another example of an unexpected outcome from a politically motivated action.

Conclusion

The demise of the Stuart line of kings had left behind a romantic Jacobite organisation which was closely intertwined with Freemasonry. The ongoing reminder of the lost line of kings was seen as a threat to the stability of the Hanoverian monarchy and so the Jacobite strands of Freemasonry were initially discouraged and then actively suppressed. The worldwide rise in Jacobite lodges was seen as a serious threat to Hanoverian security, particularly after the American and French revolutions.

The final suppression of any Jacobite origins for Freemasonry occurred with the creation of a United Grand Lodge of England, under the Duke of Sussex. Sussex purged Freemasonry of any Jacobite sympathies and completely removed or ridiculed any rituals that hinted at Stuart involvement in Freemasonry.

Once he had re-organised Freemasonry the Duke of Sussex forced himself on the Royal Society as President. While in office he reorganised the library of the Society and probably took the opportunity to make sure any incriminating Jacobite works about the Society's history were destroyed. Histories of Freemasonry, known to have been written by founders of the Society, Sir Robert Moray and Elias Ashmole, might well have disappeared at this time. Once the Royal Society had been tidied up the Duke of Sussex retired, but in the meanwhile the scientists had gained control of the Society during the political backlash to his rigged election. From that time onwards the Royal Society became a purely scientific society.

I knew from my studies of his actions within Freemasonry that Sussex had revised and changed all the ritual which linked Freemasonry to the Jacobite cause and had removed all references to the Stuart patrons of Freemasonry. It is not known when Moray's and Ashmole's *Histories of Freemasonry*, which were written at the behest of the Royal Society, disappeared. What is clear is that they were not listed in the first catalogue of the Royal Society's collection. Was the removal of these potentially embarrassing records of Jacobite Freemasonry's links with both the early Royal Society and the deposed line of Stuart kings the reason the Duke of Sussex was so keen to purge the Royal Society of non-scientific manuscripts?

As Grand Master of UGLE, Sussex wiped out all traces of Jacobite influence from the history of Freemasonry; perhaps he also wanted to be sure no clues to its Jacobite heritage were accidentally left in Freemasonry's scientific offspring, the Royal Society. If that was his intention, he almost succeeded. But strangely enough confirmation of the role of Jacobite Freemasonry in founding the Royal Society did not come from Freemasonry but from a Roman Catholic condemnation of Hanoverian dominated Freemasonry. This I will explain in the final chapter of this book.

CHAPTER 13

Sir Robert's Heritage

Freemasonry was founded to become a Counter-Church . . . Freemasonry does not destroy Churches, but prepares to replace them, thanks to the progress of ideas . . . The real inspirer, whose thought was hidden under a bushel, of the men of 1717 seems clearly to have been Thomas Spratt, Bishop of Rochester, histographer of the Royal Society, for which he thought out a programme which was to be taken up later by Anderson and transposed into the broader framework of the Grand Lodge . . . Can one go as far as to imagine some secret plan, in which Freemasonry would appear as the daughter of the Royal Society?[1]

Alex Mellor, Avocat a la Cour de Paris, writing under the Imprimatur +Patritius Carey, Vic Gen, Westonasterii, die 15th Sept 1964

IN THIS QUOTE, ALEX MELLOR claims Thomas Spratt founded Freemasonry. He is wrong to claim Spratt invented Freemasonry but he is right to see Masonic ideas in Spratt's *History*. These ideas, as we have seen, can be traced back through Sir Robert Moray, to the Freemasonry of William Schaw and James VI.

Early Scottish Freemasonry had a clear tolerance of different religious views and a farsighted dedication to studying the hidden mysteries of nature and science. These ideas can be seen in this passage Mellor chooses to quote from Spratt:

> *As for what belongs to the members themselves, that are to constitute the Society, it is to be noted that they have freely admitted men of different religions, countries, and professions of life. This they were obliged to do, or else they would have come*

> *far short of the largeness of their own declarations. For they openly profess not to lay the foundations of an English, Scotch, Irish, Popish, or Protest: philosophy, but a philosophy for mankind.*[2]

Mellor comments on how similar this statement of Spratt's, published in 1667, is to a statement about the inclusive nature of Freemasonry to be found in Anderson's Book of Constitutions, of 1723. But Anderson, elsewhere in this first Official Book published by Freemasonry about itself, talks of 'our great Master Mason Inigo Jones'; he styles James VI(I) and Charles I 'Masons' and goes into even greater detail about the period of the Restoration, saying:

> *After the Wars were over and the Royal Family restor'd, true Masonry was likewise restor'd; . . . King Charles II, was an accepted Free-Mason . . . But in the reign of his Brother King James II, the Lodges of Freemasons in London much dwindled into Ignorance, by not being duly frequented and cultivated.*[3]

Anderson clearly believed that Freemasonry existed from the reign of James VI(I). So Mellor could not be correct in saying that, in these writings of Spratt, he had found the beginnings of the Masonic philosophy. I believe that what he had noticed was the Freemasonry espoused by both Sir Robert Moray and Charles II.

My quest to understand the very odd details of the start of the Royal Society was nearly complete. I had most of the pieces of the jigsaw that explains how twelve men, drawn in equal numbers from opposite sides of a bloody and protracted civil war met to form a revolutionary scientific institution. I now knew that it was not just a happy accident but also the result of a careful plan, hatched by a canny Scotsman, Sir Robert Moray.

Now I could try to sum up the motives driving the founders

of the Royal Society and at last appreciate the full extent and richness of the role played by Sir Robert Moray.

I had an answer to my question 'what inspired an unlikely group of refugees from both sides of the Civil War to meet; form the world's oldest and most respected scientific society; and then go on to develop the tools of modern science?' It was not just chance but a carefully managed plan, organised by a skilful diplomat and trained spy. I now felt I could piece together the story of how it happened with just a little speculation to fill in the few gaps.

The Full Story of Sir Robert Moray

Sir Robert Moray, as a boy, had been fascinated by civil engineering. He grew up to become first an extremely capable soldier and then a politician. Moray had a natural interest in science and engineering. He developed this interest through a career as a Quartermaster-General in the French and Scottish Armies. While serving in the Army he became a Freemason, and found that the ideas and philosophy of Freemasonry complemented his love of science and met a need for spiritual fulfilment which he had not found in conventional religion. It also encouraged his innate love of symbolism. His studies of Freemasonry helped him think things through for himself and he developed distinct ideas which he stood by throughout his life. This self-sufficiency often provoked his enemies but he had learned from Freemasonry to be cautious in his responses. He once wrote of himself:

> *I have been reported to be writing against Scripture, an Atheist, a Magician or Necromancer, and a malignant for ought I know by half a kingdom.*[4]

It did not seem to bother him greatly! Nor did it seem to worry Charles II. Stevenson says of this relationship:

> *[Moray] was lucky to find in the cynical Charles II a king indifferent to religion who let him go his own way, remarking teasingly that he believed Moray was head of his own church.*[5]

However, this comfortable relationship with the Stuart kings developed towards the end of his life. As a young soldier Moray showed a talent for manipulation and espionage, and a weakness for the glamour of the French Court, which worked against Charles I.

While an agent for the French he was active in the events leading up to the impeachment of Charles. Moray used his membership of the Lodge of Edinburgh, which had among its members many of the Scottish courtiers of Charles I, to improve his network of contacts. The Stuarts had been involved with Freemasonry since 1601, when James VI(I) had become a Freemason and it is very likely that Charles I had also been initiated into Freemasonry, although there is no record of an Initiation.

As a member of the Scots Guard of the king of France, Moray was adopted by Cardinal Richelieu to spy on the English. He seems to have carried out this role with enough relish to be worthy of his personal motto: 'To be, rather than to seem'.

Moray clearly worked in his own interests prior to 1650. After the death of his spymaster, Cardinal Richelieu, Moray carried the news of Richelieu's death to Charles I at Oxford. Moray's connections with the Freemasons of Charles's Scottish Court may have persuaded the king that he could be trusted. In 1642, Charles knighted him in order to give him sufficient status to act as the British king's messenger to the king of France.

When Moray, who was still a serving soldier in the elite Scots Guard of Louis XIII, returned to France and delivered the message of Charles I asking for support, he was promoted for his efforts. He then went on active service in Bavaria where he was unlucky enough to be captured and imprisoned. Louis XIII then died and Cardinal Mazarin seized power over France. The new king, Louis XIV was too young to rule. Mazarin was not interested in Moray and so did not ransom him, instead leaving him to languish in prison. Moray was only ransomed when Mazarin saw a chance to use him in the bargaining between Charles and his English Parliament. Moray's Masonic connections with the Covenantors were the key to his importance. It seems Mazarin only bought him out of prison to use him again as an *agent provocateur* against Charles.

Sir Robert came close to persuading Charles I to flee to France, where he would have become a very useful pawn for Mazarin. If Charles had not lost his nerve at the last moment then Moray might have so compromised the Stuart line that Cromwell would have created an enduring English Republic. However, Charles backed out of the arrangement and was subsequently tried for, and found guilty of, treason.

After the execution of Charles I, Moray left the French Army and returned to Edinburgh to renew his contacts with the Edinburgh Lodge. After his marriage Moray seemed to become less mercenary. Up to that time his talents had been for sale and France paid him well. After his short, tragic marriage to Sophia Lindsey he seems to have developed a greater tendency towards loyalty. He got to know Charles II at a time when the young man was under tremendous religious and political pressure from the Presbyterians and Moray warmed to the Prince. From then on he seems to have used all his undoubted military and political skills to support the new Stuart king.

From the time he first met Charles II, while assisting in the negotiations for the Coronation, at Scoon, Moray became a staunch supporter of the Stuart cause. He probably attended the crowning of Charles II at Scoon Palace. And it may also have been Moray who organised an initiation for Charles into the Craft, when the king slipped away to meet with the Masonic dominated Engagers of the Earl of Lauderdale.

After the death of his young wife, Moray became very close to Charles II and was actively involved in organising an uprising on his behalf in the Highlands. When Lord Glencairn falsely accused Moray of plotting against the young king, Moray made a peculiar Masonic appeal to Charles to protest his innocence. After receiving this letter Charles spoke up in his defence. His choice of words when appealing to the king drew attention to his ongoing involvement with Freemasonry.

Professor Stevenson said of Moray's plea, 'Your Majesty may, do with me as a Master Builder doth with his materiall':

> *Is it merely that Moray, searching for a novel and forceful way of emphasising the extent of his willingness to submit comes up with a heartfelt Masonic metaphor arising from his connection with the Craft? Or did he intend Charles to recognise the Masonic reference and, as a result of the values he knew were associated with the Craft, take this as an indication of Moray's honesty and loyalty? The latter interpretation may seem unlikely, but it is nonetheless possible.*[6]

Later Moray worked for Charles in the Highlands, against the Roundheads, and he remained loyal even after being imprisoned and falsely accused of plotting to kill the king. Once his name had been cleared Moray used his influence in France to help aid the king. Charles fled to France, to join his mother, after the Roundhead invasion of Scotland and the failure of Glencairn's rising. Moray later became part of Charles's Court in Paris and then moved with the king to Bruges.

After the death of Cromwell it became clear that Charles II would be restored to the throne of England. Charles was close to his sister, who was married to the Duke of Orange and from her he knew that the naval war with the Dutch, started by Cromwell, was likely to flare up again. Moray was either asked, or volunteered, to use his Masonic contacts to gain as much military information about the intentions of the other Dutch states as he could. With this mission in mind he went to Maastricht, where he collected political and military information about the intentions of the Dutch. He used his Freemasonic links to join the local Masons and on the basis of this acceptance became a citizen of Maastricht. The purpose of Moray's spying missions was to size up the Dutch threat and he then returned to Paris to assess the likely French response before finally joining the king in London.

Once Charles was settled back in Whitehall, Moray returned to join him. When he arrived in London he was greeted as an old friend, 'the king gripping and shaking his hand' like a brother, and he was given private apartments in the Palace of Whitehall with regular access to the king. However, Moray had returned with the worrying news that the Dutch navy outclassed Charles's fleet and that a resumption of the naval war was extremely likely. Charles had no money and very little expertise to call on to improve his navy. He had a great enthusiasm for naval matters but no resources. He must have discussed the problem at great length with Moray, when Sir Robert reported back with his intelligence from the Netherlands. What could be done, without any naval experts or the money hire them?

It was Moray who came up with an inspired solution. He renewed his Masonic contacts in and around London, probably with the idea of finding out just who was involved in studying science. Within weeks Moray had made contact with the Masonic groups who were now supporting the 'poor and distressed' brethren who had been thrown out of academic office by the return of a Royalist Government.

He quickly discovered that the main centre for Freemasonry, in Restoration London, was at Gresham College. At this public college, which Sir Thomas Gresham had set up to support his Masonic ideals of study, Moray found the answer to Charles's dilemma. When the king had returned to England many of the Parliamentarian scientists had been thrown out of their university posts and were struggling to survive. An important group was based at Gresham College, mainly surviving on the small stipends the College paid to either them or their friends. Here was a pool of expertise in naval technology that could be tapped. But these 'scientists' were all politically out of favour and extremely short of money. Charles could not afford to pay them.

Moray, however, was resourceful. He had old contacts with the Masonic Scottish nobles and knew many wealthy gentlemen Masons. These Freemasons were only amateurs in the study of science but they had money and influence. Moray saw a way of harnessing these two groups and persuading them to work together for the good of their king and country. He was aware of the ideas behind Richelieu's *Académie Française* and saw that he could use his Masonic contacts to build something similar in London to solve the problems of Charles's navy.

Moray brought together Royalists with money and Parliamentarians with scientific skills, in order to set up a self-funded group to help solve the pressing problems of sorting out the navy. It is clear that Moray was afraid of another war with the Dutch and he realised that their shipbuilding skills were far in advance of those to be found in England at the time. His solution touched the imagination of the newly restored kingdom. He used the interest in science, which was shared by all Freemasons, as a basis for a new Society to focus the application of science on the problems of defence.

Sir Robert encouraged his friends and contacts to attend the weekly lecture, held by one of the bright stars of the Parliamentary scientists, Christopher Wren. It would seem that two of the members of that first meeting were definitely not Masons, Christopher Wren and Robert Boyle. They are recorded as being at the first meeting but have also been added to the list of members drawn up at the meeting to be the first to be invited to join. This omission can only be explained if they had left before Moray and his Brother Masons got down to the detailed discussion of setting up a new society to study the Masonic objectives of the hidden mysteries of nature and science. Although it seems that Wren may have become a Freemason at a later date, Robert Boyle never joined the Craft, as he would not take an oath under any circumstances.

To make his idea work Moray took from Freemasonry the injunction not to speak about religion or politics within the meetings. And he drew funds by appealing to the charity of those who could afford it, so enabling able but poor men to carry out experiments.

Moray won the confidence of the Parliamentary Masons when he made sure that their deposed leader, John Wilkins, took the chair of that first meeting. Wilkins had been extremely close to Cromwell and his family. By rehabilitating him with the king, Moray showed the other Parliamentary scientists that they were all equal in the new Masonically inspired scientific body he was creating. He laid his ground carefully and, despite the king's busy schedule, Moray reported back to the group, within a week, that they would receive a Royal Charter.

For the first two years he drove and chivvied the group towards his vision of a new scientific Navy. He was satirised as this verse about him shows:

The Prime Virtuoso hath undertaken
Through all the Experiments to run
Of that learned man, Sir Francis Bacon
Shewing which can, which can't be done.[7]

Moray made sure that most of the scientists, among these very first members, had an interest in subjects that mattered to the navy. He encouraged ship designers, navigation experts and weapons specialists to contribute to the early work. At first he ensured that he chaired the majority of the meetings himself, to establish a structured form of meeting. Moray followed an agenda and kept minutes, ways of working he had learned from the Schaw Freemasonic Lodges of Scotland. The two basic rules he laid down were: all men were welcome to join, irrespective of politics, race or religion; and during meetings only scientific matters were to be discussed, religion and politics were forbidden. Small wonder that Alex Mellor (quoted at the beginning of this chapter) found a flavour of Freemasonry in the writings of Thomas Spratt. Spratt wrote his *History* under the editorial direction of Moray. But Spratt did not invent the principles of Freemasonry; he learned them from the Masonic practices of the early Royal Society.

Moray succeeded in creating something far greater than he had ever dreamed of. As the Society was developing, it was taking on a life of its own. It soon started to separate from its Masonic roots. Moray groomed others to take over the day-to-day tasks of running the meetings and devoted himself to drawing up a charter for his brainchild. As the society grew it took in many more non-Masons.

When the First Charter was delivered Moray stood back, putting forward the Naval enthusiast, Lord Brouncker as the First President, hoping that the Society would now continue under its own momentum. Perhaps he now hoped to spend more time working on the history of Masonry that he had started to write and encouraging the free exchange of information by his proposed 'Transactions'. The Society did continue to find its own way and it struck out for a more independent role.

The first sign that Moray's society was developing into something more than a specialised Masonic Committee to support the King came when he proudly presented the First Charter to his Royal Society. The fellows did not like the title, which perhaps was too much of an indication of Moray's intent. They wanted a title that linked them with science, not just with Royalty. Its members insisted on a title which made them more than just a 'Society for Supporting the King'; they became a Society for the pursuit of knowledge, which was patronised by the King. However, the principle Moray had established of mixing together wealthy amateurs to provide the funds, and less wealthy scientists to do the work of experimentation, proved to be sound for nearly two hundred years.

The Masonic philosophy inherited by the new Society led to the nurturing of the most important scientific developments of all time. The problems faced by Charles's navy were the problems of understanding the Universe. By developing techniques to aid navigation the founders of the Royal Society created techniques and technology which enabled their members to study the stars. The policy of carrying out flamboyant demonstrations spread the ideas of science to the more influential layers of society. By using the microscope to investigate minute creatures to amuse the nobility, the science of biology was discovered. Finally the policy of publishing the results of studies and experiments increased the rate of innovation. In less than twenty years the study of the stars had moved from the lore of astrology to the practical application of Newton's Laws to predict the return of Halley's Comet. It is a curious thought that the first edition of *Old Moore's Almanack* was published just seven years before Newton's study of the heavens transformed Francis Moore's science of astrology into mere superstition. The newly formed Royal Society was a potent package which took a lively group of thinkers and gave them funding, encouragement and a means of sharing knowledge. Without the change in attitude to the study of the skies, which the Royal Society had achieved, Newton might never have been published. Less than a generation earlier, while Bacon was writing of his Solomon's House, Galileo was being persecuted by the Church for daring to suggest the Earth revolves around the sun!

All Freemasons today recite the formal statement of the Galilean heresy which forms part of the test questions of the Fellowcraft Degree. Is this a permanent memorial to the work of Bro Sir Robert Moray in putting into practice his Masonic Oath to 'study the hidden secrets of Nature and Science in Order to better know his Maker'?

Despite the evidence of his actions I find it hard to believe that Sir Robert really set out to create the world's premier Scientific Society on 28 November 1660. He probably only expected the group to solve the military problems Charles could not afford to tackle. However, he had used the Masonic principles of equality and the study of science to create a tremendous living force. His group was free from the shackles of religious dogma and had a unique structure for its time. Whether by accident or design, he had used three of the most powerful ideas of Scottish Freemasonry but he had applied them to the development of technology. These were the ideas he took from Freemasonry:

1. That the study of the works of nature can lead to an understanding of the underlying plan of God, i.e. that there is an underlying order to the laws of nature that can be determined by observation and experiment. This idea led directly to the work of Newton.

2. That all men are equal. If they come together to discuss learning, and forbid discussion of religion and politics they will be able to cooperate. In this way he encouraged scientists, who had been strong supporters of Parliament, to sit down and meet with wealthy Royalists, who in turn

helped fund their work and assist their rehabilitation into Restoration society. This concentration on experimental science to the exclusion of all distractions aided the Royal Society in becoming a major force in creating our modern scientific age.

3. That for officers and Presidents to have true power, they must be elected by, and have the support of, the members they rule. As William Schaw had decreed sixty years earlier, so Moray built into the charters of the Society, that the Fellows should elect their own leaders so that they would be loyal to them.

These principles proved to be a sound foundation for building a scientific institution. Moray's fourth principle, that wealthy amateurs could be brought into the Society to fund less wealthy scientists, was an idea that only lasted until the Presidency of the Duke of Sussex. Since Sussex's time the Royal Society has limited its members to scientists of worldwide renown.

The Masonic roots of the Royal Society have been suggested by many writers, some such as Alex Mellor even going so far as to say that the founders of the Royal Society gave birth to Freemasonry, but I believe I have shown that this is not the case. The Royal Society is the child of Freemasonry.

So now I have put together my version of the complete story behind the unlikely success of the Royal Society. It was founded by an astute, politically motivated, Freemason. Its purpose was to solve a short-term crisis in military technology for a run-down navy. Sir Robert Moray took the structure and philosophy of Scottish Freemasonry and used it to build a totally new type of organisation. This soon outgrew Moray's limited aims and drew up for itself a much wider agenda, and while doing so it took the best of Moray's ideas and applied them to its own choice of problems.

Its new attitudes to knowledge and the study of the hidden mysteries of nature and science led to the successful study of physics and the theories of Newton. Natural Philosophy became a predictive science and superstition turned into technology. We owe our modern society, and its many wonderful scientific innovations, to the accidental success of Sir Robert Moray. He saw the wisdom of the Masonic teachings, which had inspired him; he used the Schaw Lodge System and the workings of Masonic harmony to bring together the opposing sides of the great Civil War; and he provided a structure that enabled science to break free of the superstitious cage of religion.

There is an unwritten rule of politics known as the Law of Unexpected Consequences. It says that no matter how carefully you analyse a complex situation you will not be able to foresee all the possible outcomes of your actions. This is certainly true of the founders of the Royal Society. This small group of Freemasons probably only expected to solve some of the problems of naval technology or perhaps get back some of their lost position in society. What they did was much greater and they created a system that developed the possibilities of a vast increase in human wellbeing, more than any other in recorded history.

Scientific method started with the work of the Royal Society and it in turn was inspired by the teaching of Scottish Freemasonry. Let us not forget how much we all owe to our Antient Brethren, Bro William Schaw, Bro Sir Robert Moray, Bro the Revd John Wilkins and even Bro His Royal Highness Charles II. Later political events may well have made it expedient for the Hanoverian Monarchy to forget the debt our society has to Scottish Jacobite Freemasonry, and the United Grand Lodge of England may prefer to be coy about its Scottish roots, but hasn't enough time now passed for the threat of a Jacobite revival to be discounted?

Surely now we can freely celebrate the true story of the Masonic birth of modern Science and honour the memory of Brother Sir Robert Moray, the man who conceived the Royal Society, nurtured it through nine months of early presidencies and finally brought it to birth with its founding charters.

POSTSCRIPT

Life, the Universe and a Theory of Everything

THIS BOOK HAS BEEN A PERSONAL quest to understand the unlikely circumstance of the birth of the Royal Society. It was fuelled by my interest in science and directed by my disbelief in over-simple explanations of its beginnings. A 'Big Bang' rocked science some three hundred years ago and the echoes of the explosion still shake modern society. The cause of this eruption was the world's first modern scientific society. It was a society that unshackled scientists from superstition and left them free to build modern technology. This revolution lay behind all the subjects I wanted to study when I decided, as a schoolboy, that I wanted to be a scientist.

To become a scientist I had to study maths. And while studying for an 'A Level' in Applied Maths I was introduced to the men who had dominated the mathematical thinking of science for hundreds of years. Most important of these was

Isaac Newton, first a Fellow and later President of the Royal Society.

As an apprentice to the trade of applied mathematics I studied Newton's three laws of motion in a multitude of different circumstances. The laws themselves are very simple. Here they are as I first learned them:

1. Every body continues in a state of rest or of uniform motion in a straight line, except in so far as it be compelled to change that state by external impressed forces.

2. Change of momentum per unit time is proportional to the impressed force, and takes place in the direction of the straight line in which the force acts.

3. To every action there is always an equal and opposite reaction; or, the mutual actions of any two bodies are always equal and oppositely directed.

My textbook then went on to add, 'For the first two of these laws we owe much to Galileo.'[1] Galileo was the other giant of scientific method, whose example my tutors urged me to follow.

These basic laws were first observed in the experiments of Galileo. But it was Newton who gave them the mathematical form I was taught. I learned to calculate the result of all sorts of bumps and crashes between particles. The impact of elastic and non-elastic bodies; the movements of pendulums and springs; where to find centres of gravity; predicting motion in a circle; the effects of friction; the details of mechanical advantage; all surrendered their mystery to the analytical power of Newton's simple laws. And when I looked up towards the Moon, inspired by the excitement of President Kennedy's announcement that NASA was going to put a man there before the end of the decade, it was Newton who helped me comprehend the rocket science that controlled the orbits of the astronauts.

Newton's law of gravitation was just as simple to write out and just as difficult to learn how to apply:

> *Every particle of matter attracts every other particle of matter with a force which varies directly as the product of the masses of the particles and inversely as the square of the distance between them.*

This law could explain almost every visible movement of the stars over my head. For three hundred years it remained unchallenged as the best way of describing the universe. Newton wrote a best-selling book about his work; he called it 'Mathematical Principles of Natural Philosophy' *(Principia Mathematica*). It remains in print to this day, although current editions have been translated into English rather than the Latin Newton used to write it.

Newton saw his work as a means of studying the will of God. In his own words:

> *This most beautiful system of the sun, planets, and comets, could only proceed from the counsel and dominion of an intelligent and powerful Being . . . This Being governs all things, not as the soul of the world, but as Lord over all and on account of His domination He is wont to be called Lord God.*[2]

Newton was an experimental physicist, who saw the study of nature as a way of understanding the Mind of God. He was convinced that a total understanding of everything was possible. He closes his book with these words:

> *And now we might add something concerning a certain most subtle spirit which pervades and lies in all gross bodies; by the*

> *force and action of which spirit the particles of bodies attract one another at near distances, and cohere, if contiguous; and electric bodies operate to greater distances, as well repelling as attracting the neighbouring corpuscles; and light is emitted, reflected, refracted, inflected and heats bodies; and all sensation is excited, and the members of animal bodies move at the command of the will, namely, by the vibration of this spirit, mutually propagated along the solid filaments of the nerves from the outward organs of sense to the brain, and from the brain into the muscles. But these are things that cannot be explained in few words, nor are we furnished with that sufficiency of experiments which is required to an accurate determination and demonstration of the laws by which this electric and elastic spirit operates.*[3]

Newton is confident that his sums can solve every puzzle of life and the universe. However, as I progressed through a degree in Electronics and went on to work towards my PhD in Solid State Physics I learned he was wrong. His elegant, predictable model of the universe didn't explain the physics of very small particles I was studying.

The predictable collisions between billiard balls, which Newton's laws had allowed me to analyse, gave way to the erratic outcomes of quantum theory. I had to face the difficult concept that there is a physical limit to my ability to see what is going on. Heisenberg's Uncertainty principle insisted that there is a basic lumpiness in the structure of the universe that will stop me from ever seeing much fine detail. If I knew where a particle was, the veil of Planck's constant stopped me knowing how much energy it had. I had to learn a new set of tools for small events that relied entirely on statistical analysis. I had to accept that I could only predict the behaviour of crowds of particles and never really know the detailed movement of any single one.

Newton founded a confident view of science where every event in the universe was foreseeable. He believed in a clockwork Universe. If he collected enough information about the past movement of any object he knew he could accurately predict what would happen to it in the future. True, the sums might be complicated and difficult, but they were always possible. He even suggested this way of thinking would extend to the actions of humans. The world of quantum physics has no such certainty. It broke apart Newton's continuous connected predictable space into a welter of separate and discrete packets of energy all arrogantly defying measurement.

When I was a student, Newton's supremacy had already been undermined by the work of Albert Einstein. He had shown that Newton's assumption that time was the same everywhere in the universe could not be true. This thought led Einstein on to understand the link between energy and mass that underlies how gravity works. Newton had never understood how gravity worked, he admitted that he could describe and predict its effects but had no explanation of why particles attracted each other. Newton says:

> *To us it is enough that gravity does really exist, and acts according to the laws which we have explained, and abundantly serves to account for all the motions of the celestial bodies, and of our sea . . . we have explained the phenomena of the heavens and of our sea by the power of gravity, but have not yet assigned the cause of this power . . . I have not been able to discover the cause of these properties of gravity from phenomena, and I frame no hypotheses.*[4]

Einstein framed such a hypothesis saying that gravity was a curvature of space. When this was confirmed by observing starlight bending as it passes close to the sun, then a whole new field of physics opened up. This work led to the development of nuclear power. But now physics had split into two major viewpoints. A large-scale view of the universe, which said that space and time were continuous and could be measured to any degree of fine detail; and the small-scale world of atomic physics that said everything came in small packets that wouldn't stand still to be measured and had to be understood statistically. As a young postdoctoral researcher in semiconducting integrated circuits I lived with this split view. I used Fermi-Dirac statistics to think about the orbits of electrons in crystals. But, in my hobby of astronomy, I used Newton's laws to calculate orbits and particle motion in space. Then I came across the work of yet another Fellow of the Royal Society which for the first time showed a way to join the disparate parts of my schizophrenic world into a coherent whole.

In the mid-1970s Manchester University's Department of Astronomy ran a series of extramural courses about new developments in astro-physics. At the time I was working at the laboratories of Ferranti Semiconductors, in Manchester, and a colleague, sharing my hobby interest in astronomy mentioned the lectures to me. Together we enrolled on the course. This was where I first heard about a scientist who was trying to reconcile the classical ideas of continuous space with the inherent disconnectedness of the quantum theory I used in my daily work.

On that course I learned about black holes, objects so massive that not even light can escape from them and how quantum theory would allow them to be seen via the Hawking radiation they must emit. At this point I rushed out to buy Hawking's new book *The Large Scale Structure of Space Time.*[5] About three pages into it I decided I was not mathematically competent to become an astro-physicist and went back to simple things like electrons that instantaneously jumped across energy barriers when I forced them to adopt a known energy level. After all, I rationalised, the scientists I worked with still used Newton's laws to calculate the orbits of the communication satellites that used the integrated circuits I made.

I did, however, continue to be fascinated with Hawking's thoughts and when he wrote a popular guide to his work I found it immensely enjoyable. A well-thumbed copy of his *Brief History of Time* sits along-side my still pristine edition of *The Large Scale Structure of Space Time* on the bookshelves of my study.

Hawking has become an internationally celebrated scientist. It is, however, interesting to note that his first serious recognition, as an important contributor to scientific thought, came from the Royal Society. In March 1974, at the age of thirty-two, and before the official publication of his discovery of Hawking Radiation, he was made a Fellow. For once the Royal Society broke with tradition. Instead of the new Fellow walking up to the podium to sign the roll of honour, the President, Sir Alan Hodgkin, brought the book to the candidate to sign. As Stephen Hawking suffers from a serious muscle-consuming disease he is the person who took the longest time to sign his name in the entire history of the Society. The assembled scientists stood in complete silence while he painstakingly argued with his wasted muscles, insisting they must move the pen to form his signature. When he finally completed the protracted labour of signing, the watching company broke into enthusiastic and heartfelt applause at the determination of this man who never gives in to his disability.

Hawking has said on many occasions since his investiture as an FRS that it is the proudest moment of his career. And this is a man who has won almost every serious scientific prize it is possible to win. Membership of the Royal Society has become the highest honour a British scientist can hope to gain. Hawking went on to follow Newton into the Lucasian Chair of Mathematics at Cambridge, though as Hawking himself said in one of his early professorial lectures, 'when Newton occupied this chair it didn't have wheels on it!' It is impossible and unnecessary to pity Stephen Hawking just because he was unlucky enough to contract ALS. But it's very easy to be impressed by the scope of his scientific vision and inspired by the implications of his ideas.

Galileo, Newton and Hawking. These three great scientists are strangely linked! The year Galileo died Isaac Newton was born. Then on the three hundredth anniversary of Galileo's death, 8 January 1942 Hawking was born, coincidentally three hundred years after the birth of Newton.

When Hawking was born, Newton's place in cosmology had not been usurped for 250 years. Einstein's Theory of Relativity still kept to Newton's assumption that space and time are continuous and can be measured just as accurately as you wish. Einstein was never happy with the religious consequences of quantum theory. He is reported to have said to quantum physicist Neils Bohr, 'God does not play dice with the Universe.' Bohr's famous reply was, 'Albert, it is not your place to tell God what he can and can not do'! Hawking has challenged all this and has made the first steps to integrate quantum theory with relativity and thermodynamics. This thinking has led him to ask the sort of questions that were once the sole preserve of the Church. How and when was the Universe created? And does it need a creator to make it work? Questions Newton took for granted as matters of faith.

Galileo was forced to moderate his science to fit the view of the Church that as God's creation the earth had to be at the centre of the Universe. Newton created an infinitely large deterministic universe that was a tribute to the organisational capabilities of the creator God who ruled it. Hawking feels able to ask, 'is there a need for God in a theory of the creation of the Universe' and answer 'no'.

Hawking has come a long way from Galileo. Galileo was a ground-breaking scientist, who also wrote a best-selling book about his work with the snappy title *Dialogue Concerning the*

Two Chief World Systems, which was published in 1632. This book eventually led to Galileo's public renunciation of his life's work before the Inquisition.

If the Inquisitors-General of the Holy Catholic and Apostolic Church who 'corrected' Galileo had been exposed to Stephen Hawking's view of the role of God in the Universe then they would probably have had Stephen burned at the stake! ('Does the Universe need a creator, and if so, does he have any other effect on the universe? And who created him?'[6]) It seems highly unlikely a man with Hawking's glorious stubbornness would have willingly submitted to the humiliation that Galileo did.

Yet by the time Newton published his own best-selling book, in 1686, he fell quite at ease to speculate at great length about the role of God as the general manager of a heliocentric solar system. Newton felt quite secure making the statement:

> *The true God is a living, intelligent and powerful Being; and, from his other perfections, he is supreme, or most perfect. He is eternal and infinite, omnipotent and omniscient, his duration reaches from eternity to eternity; his presence from infinity to infinity; he is not duration or space but he endures and is present.*[7]

The whole climate of the church's censorship of science had changed in the fifty years between the publication of Galileo's work and that of Newton. Newton laid the foundations of modern physics. Without his work we would have no space science, no electronics industry, no mechanical devices and no understanding of the universe we live in.

The change in climate that freed Newton to think his great thoughts was brought about by the Royal Society and its carefully thought out rules of conduct. Yet its creation, in the aftermath of the Civil War and the Restoration was incredibly unlikely. We owe it so much and yet know so little about what drove its founders to overcome the enormous difficulties of its foundation.

It matters little that Sir Robert Moray had a limited, and somewhat sordid, political aim when he created the Society. He saw a major problem for Charles II if the technical difficulties of the British Navy were not addressed, and his solution was a Royal Society. This Society did an extremely good job in the area of naval research, but it went on to be far more than Moray ever dreamed of.

The problems of navigation were also the problems of life, the universe and everything! Once scientists were freed from the shackles of politics, religious dogma and superstition they set about addressing these questions. The work of Stephen Hawking is a direct descendant of the work of Isaac Newton. The Royal Society recognised this link in making Hawking an FRS. It was a few years before Cambridge caught up and appointed Hawking to Newton's old Chair at the University. The whole population of the world has benefited from the technological improvements in living standards that science has accomplished in the three hundred and forty years since the scientists of the Royal Society first broke free from the chains of the Church's restricted worldview. A view that bound Galileo right up to his death. But the real monument to Sir Robert Moray's inspired solution to Charles's naval problems is the work of Hawking.

It is a sobering thought to realise that Stephen Hawking would not have lived long enough to carry out his pioneering work without modern technology. His ability to continue to work, despite his crippling disability, stems from the effects of so many previous FRS's who developed the science which keeps him alive and functioning. His electric wheelchair, his mechanical aids, his computers, the medical expertise of his doctors and nurses would never have been developed if the writ of the Inquisitors-General of the Holy Catholic and Apostolic

Church had not been challenged, and defeated, by the founders of the Royal Society. The work of Hawking would not have been possible if he had been forced to use only the technology and science accepted by the Church in 1632. If he had been compelled to write out *A Brief History of Time* in longhand he might still be forcing his pen to slowly form the letters and the world would be much the poorer.

Newton said of himself: 'I do not know what I may appear to the world, but to myself I seem to have been only like a boy playing on the seashore, and diverting myself in now and then finding a smoother pebble or a prettier shell than ordinary, while the great ocean of truth lay all undiscovered before me.'

Hawking propelled his electric wheelchair down onto that beach and sat looking out towards the horizon of that great ocean, then using his computer generated voice, he shared with all humanity the truths he saw. His achievements are intertwined with the achievements of the Royal Society, and the advances in technology that allow him to continue to work flow from the breakthroughs of the seventeenth century.

It somehow seems appropriate that Hawking should be the most distinguished living member of the Royal Society, the successor to Newton and a direct beneficiary of the scientific heritage of Sir Robert Moray.

APPENDIX

The Scientific Secrets of the Craft

OUTSIDERS OFTEN VIEW FREEMASONRY as a strange, secretive and spooky set-up. This is a sad state of affairs because Freemasonry is basically a moral organisation, which does a lot of charitable good work but still has a poor public face. Ill-informed attacks by the Press; by religious fanatics; and by politicians, intent on demonising it, have marginalised this eccentric society, at least in England. But Freemasonry itself has done very little, until recently, to change this impression. This appendix fills in the main technical background points about Freemasonry, for those readers who may not know much about the Craft, and explains the scientific bias underlying some of the rituals.

Freemasonry started in Scotland, so I will start by quoting how the Scots define it. The Grand Lodge of Antient Free and

Accepted Masons of Scotland, says that Freemasonry is as follows:

> *Freemasonry teaches moral lessons and self knowledge through participation in a progression of allegorical two-part plays, which are learnt by heart and are performed with each lodge. Freemasonry offers its members an approach to life which seeks to reinforce thoughtfulness for others, kindness in the community, honesty in business, courtesy in society and fairness in all things. Members are urged to regard the interests of the family as paramount but importantly Freemasonry also teaches and practises concern for people, care for the less fortunate and help for those in need. Membership is open to men of all faiths who are law-abiding, of good character and who acknowledge a belief in God. Freemasonry is a multi-racial and multi-cultural organisation. It has attracted men of goodwill from all sectors of the community into membership. There are similar organisations for women.*[1]

Freemasonry is a society that meets regularly in groups of not less than seven individuals. These are called lodges. Each lodge has a name and, since the formation of Grand Lodges after 1717, a roll number. Freemasonry is an organisation that teaches the practice of morality and charity, and it does so in a non-religious context. It is open to everyone who can express a belief in a Supreme Being. As a Freemason you will never be asked the detail of your religious convictions, as the discussion of religion and politics is expressly forbidden within a lodge. At the regular meetings, usually monthly, the lodge will carry out one of three basic ceremonies, which every Mason passes through, and if there is no ceremony to perform then a lecture will be held to improve the knowledge of the brethren.

The method of working (which is the term Freemasons use to describe how they perform their ritual plays) is based on a verbal tradition and every ceremony must be carried out word perfect and from memory. As a Freemason you soon learn a lot about the 'Art of Memoire' as the Second Schaw Statutes called the method.

Modern Freemasonry has three degrees, these being Entered Apprentice, Fellow-Craft and Master Mason. The form of these three degrees was largely fixed in 1816, soon after the amalgamation of the two Grand Lodges in England. One group was known as the Antients, the other the Moderns. These two groups were brought together by the then Prince of Wales, Grand Master of the Moderns, and the Duke of Kent, then Grand Master of the Antients, combining the two organisations under the control of their younger brother the Duke of Sussex. There has been no change in the structure of the three degrees of Craft Masonry since this time, although a great number of changes have been made to the detailed wording of the rituals.

Historian Victor Langton investigated the way in which the wording of the rituals has changed in Freemasonry and arrived at the following conclusion:

> *What happened in 1816 was that many changes to the then existing ritual were put into effect by the United Grand Lodge of England which had been formed just three years before in 1813, but as Speculative Freemasonry dates from about 1620, the various changes to the already established ritual from then until 1816 are of great interest, although what is perhaps the most surprising aspect of all is the amount of that very early ritual which would be clearly recognisable to the Freemason of today.*[2]

The First Schaw Statute had ordered all Masons 'to observe and keep the good ordinances set down before concerning the privileges and of their Craft by their predecessors of good memory'.

This instruction had been strengthened by the Second Schaw Statute of 1599 that confirmed the duties of the Wardens of Kilwinning, making them responsible for ensuring that all Fellows of the Craft of Masonry were 'to take trial of the qualification of all the Masons . . . of their art, craft, science and antient memory'.

From these statutes it is clear that particular rituals had to be followed and these rituals had to be memorised accurately. In much the same way modern Freemasonry insists on accurate recitation of the exact words of the ritual. But what rituals did these early Scottish Freemasons use? As Professor Stevenson pointed out Masonic records do not write about the rituals. This he says is because the obligation every Mason submits to has always forbidden writing the secrets. The early brethren had a much wider definition of the secrets of the Craft than the modern rulings admit and they believed the rituals to be secret. This is not the view of modern Freemasonry, which encourages openness about the Craft. Here I repeat the antient words of the Obligation of the First Degree and what they say about the secrets:

> *I further solemnly promise that I will not write those secrets , indite, carve, mark, engrave, or otherwise deliniate, or cause or suffer the same to be done by others, if in my power to prevent it on anything movable or immovable under the canopy of heaven, whereby or whereon, any letter character or figure or least trace of a letter, character or figure may become legible or intelligible to myself, or any one in the world, so that our hidden mysteries may improperly become known in or through my unworthiness.*

The modern interpretation of the secrets as the methods of identification means that these are never printed, or indicated, in ritual books. However, what has been recorded in ritual books, and published by the Grand Lodges, is no longer covered by this obligation. So, even though I am an obligated Freemason, I can and will discuss freely all the important teachings of Freemasonry that bear on the formation of the Royal Society.

The early Scottish Freemasons conferred two degrees. The first degree is given at Initiation and is known as the Entered Apprentice. The second after the completion of the apprenticeship is known as the Fellow of the Craft, or Fellowcraft. Sometimes both degrees were given in the same ceremony, as happened to Sir Robert Moray. Langton records evidence of Freemasons from the Lodge of Edinburgh demonstrating these two degrees to English Freemasons in York in 1615 and so founding the York Rite of Freemasonry.[3]

The United Grand Lodge of London, which makes many and frequent claims to be the Premier Grand Lodge of all Freemasonry, bases its claim on its descent from the Grand Lodge of London formed under Anthony Sayer in 1717. It is an interesting, if ironic, comment on the records of the Grand Lodge of London that the only written evidence of its formation and history prior to the formation of the Grand Lodges of Ireland and Scotland was authored by a Scotsman, six years after the event. This man, James Anderson, was minister of the Presbyterian Church in Swallow Street, London and a member of the Lodge of Aberdeen! Anderson also records that the then Grand Master of the Grand Lodge of London, Dr John Theophilous Desaguliers, travelled to Edinburgh in August 1721 to visit the Lodge of Edinburgh.

Desaguliers had been introduced to Freemasonry while employed by the then President of the Royal Society Isaac Newton, as an experiment demonstrator for the Society's meetings. Desaguliers started out as an employee of the Royal Society. Under the patronage of Newton he became a Fellow of the Royal Society, and he also rose to the highest rank in English Freemasonry.

Desaguliers seems to have received all his early instruction in Freemasonry from Scotland for he had built up close links with Dunfermline, the place where William Schaw is buried. On 20 August 1720 he was made a Free Honorary Burgess of Dunfermline, by his close friend Sir Peter Halkett. At the time Desaguliers was working in and around the Edinburgh district as a consultant, acting on behalf of the Royal Society, for the Edinburgh and District Water supply system[4] and was also a regular visitor to the Lodge of Edinburgh (St Mary's Chapel) No 1.

A year later, at this same Edinburgh lodge, he was instructed in 'the ceremonies of entering and passing, as far as the circumstances of the Lodge would permit'. Masonic historian Dudley Wright felt obliged to justify this embarrassing fact by adding a comment, as he noted it:

> *... the Doctor [Desaguliers] confined himself to the two lessor degrees ... [because] it was not till 1722–1723 that the English regulation restricting the conferring of the Third Degree to Grand Lodge was repealed.*[5]

The regulation Wright is referring to stemmed from an excess of what I can only describe as control freakery on the part of the Grand Lodge of London. After its arbitrary formation in 1717 it took upon itself a right to control any other lodges by issuing warrants to operate under the regulations of this newly self-styled Grand Lodge. These regulations insisted that to become a Master of a subordinate lodge the Freemason must have worked the ritual known as the Master's Part. Masonic historian, Victor Langton comments:

> *the Higher Degree or Master's Part could only be conferred by Grand Lodge. This change had the effect of enabling Grand*

> *Lodge to withhold the Chair of any lodge from anyone of whom it did not approve by simply refusing to confer upon them the required Higher Degree, for it clearly said in the Book of Constitutions that the Master of a lodge must be a Fellowcraft. i.e. must have taken that Higher Degree, now conferred by Grand Lodge alone.*[6]

From all the accounts I have reviewed, this Master's Part seems to be the final part of the ritual the Scottish lodges used to make a Mason a Fellow of the Craft. By controlling the award of this degree and insisting that only they had the power to confer it, the Grand Lodge of London could decide who would became masters of its subordinate lodges.[7] This practice seemed designed to centralise control of Freemasonry in the Officers of the Grand Lodge of London. As this was in direct conflict with the letter and spirit of the Schaw Statutes, that the Master 'be elected each year to have charge over every lodge', not all Freemasons accepted this rule. Instead they created a new degree using part of the Entered Apprentice Ritual and the first part of the Fellowcraft ritual. This new degree became what is now called the Second Degree. Today the title awarded with the conferment of the Second Degree is Fellowcraft. The Book of Constitutions said that a Fellowcraft could become a Master of a lodge without having to conform to a choice imposed by the self-important Grand Lodge of London. By 1723 this attempt at controlling the conferment of the Master's Part, had collapsed into failure. But, as a direct result of the refusal of other lodges to accept this attempted seizure of authority by the Grand Lodge of London, the two degrees that Desaguliers had studied in Edinburgh were split into the present day three.

While looking more closely at the motives of the founders of the Royal Society I investigated the forms of the ritual that had been used by Freemasons in the early seventeenth century. This is not as simple as looking at a modern ritual book. When I had co-authored other books on early Freemasonry I had found that the rituals used in English Freemasonry had been greatly edited and changed. This had happened both in the mid-twentieth and in the early nineteenth century. There had also been major changes soon after the formation of the United Grand Lodge of England in 1813, so I needed to think where I was likely to find versions of the Freemasonic ritual which preserved the older Scottish elements of the content.

The Oldest Rituals of Freemasonry

For many years I have been collecting as many versions of early rituals as I have been able to find, from as many different sources as possible. I have got used to plain brown envelopes containing tatty old ritual books being passed quietly to me when I visit Freemason's lodges. By comparing these rituals I have been able to build up a reasonable picture of the elements that are common across the majority of rituals. These elements are statistically the most likely to have survived from the oldest sources.

What is very clear is that the Scottish lodges have always been far less willing to change their ritual. As a result they have often preserved a form of words that is likely to be nearer to the rituals used by James VI, Sir Robert Moray and Alexander Seton. Even they, however, have in later years adopted the English system with its three degrees.

Masonic historian A E Waite, writing in the late nineteenth century, said that the original rituals of the Freemasons had been greatly changed after the appointment of the Duke of Sussex as the First Grand Master of the United Grand Lodge of England in 1813.[8] To study what the rituals had taught about the hidden mysteries of nature and science in the early seventeenth century it was clear that I would need to look to the ritual of the older Scottish lodges.

Professor David Stevenson had looked in detail at the oldest evidence for rituals of identification and initiation in Scottish Freemasonry. All these references talk of a Mason's Word. As Stevenson comments:

> *Scattered references to the [Mason's] Word occur from the 1630s onwards and through them something can be discerned of how outsiders perceived the masons and their rumoured secrets. Surveying these references thus takes on something of the character of a progressive revelation of what was known of the esoteric side of the Craft. Surprisingly, this handful of references in non-masonic sources is not accompanied by similar references to the Masonic lodges, suggesting that convention among the masons dictated that lodges should not be mentioned to outsiders, but that it was permissible (or gradually became permissible) to intrigue the un-initiated by referring to the existence of the Word – though of course without revealing its secrets.*[9]

The Stuart links to Scottish Freemasonry were also important. I had been told that Alexander Seton (first Earl of Dunfermline) had been a member of the Lodge of Aberdeen and I also knew that in 1679 his son Charles (second Earl of Dunfermline) had been recorded as one of the authorities of the lodge in a specially produced Mark book that is kept at the Masonic Hall in Aberdeen. This same Charles Seton had been one of the Scottish Royalists who fled to France with Charles II and had returned with him when he was crowned King of Scots in Scoon, on the first day of 1651.

The laws and statutes of the Lodge of Aberdeen, recorded in 1670, say:

> *Wee Maister Meassones and entered pretises all of ws wnder subscyvers doe heir protest and vowe as hitherto wee have done*

at our entrie, when we receaved the benefit of the Measson word, that wee shall owne this honourable lodge at all occassions except those who can give ane Lawful excuse of sicknes or out of town.

So the Lodge of Aberdeen has written evidence that it was using Masonic ritual during the reign of Charles II.[10] It is also the lodge to which one of Charles II's courtiers belonged during his exile and has a reputation for taking good care of its ritual. So all in all, Aberdeen looked a good place to go if I wanted to study a form of ritual that was likely to have preserved the elements of the Craft which had motivated some of the founders of the Royal Society.

The Masonic Temple at Aberdeen is a magnificent purpose-built granite building that houses not just the Lodge of Aberdeen No 1 [(3)] on the roll of the Grand of Scotland, but also nine other lodges. It is also home to the Preceptories and Priories of St George Aboyne and Aberdeen Military.

During my tour of the building and its Masonic treasures I was allowed to sit in a Masonic Master's Chair that has been in the ownership of the lodge since the seventeenth century. The date 1640 is carved on the back along with various Masonic symbols. I couldn't help wondering if Charles Seton had also sat in the same carved oak chair.

Later I compared the Aberdeen ritual with my own collections, looking for the most common elements of the Second Degree. I have put together a composite of the exposition of the principles of the degree, which are delivered to the newly made Mason by a more senior member of the lodge. I pictured the Earl of Dunfermline at his Initiation, imagining how he must have felt when he heard them for the first time:

Brother Seton, having passed through the ceremony of your initiation allow me to congratulate you on being made a member of our antient and honourable Society. Antient no doubt it is, having subsisted from time immemorial; and honourable it must be acknowledged to be as by a natural tendency it tends to make all those honourable who are obedient to its precepts; indeed no institution can boast a more solid foundation than that on which Freemasonry rests – the practice of every moral and social virtue. To so high an eminence has its virtue been advanced, that in every age Monarchs themselves have been the promoters of our Art, and have not thought it beneath their dignity to patronise our mysteries, exchange for a while the Sceptre for the Trowel, and join in our assemblies.

As a Freemason I would first recommend to your most serious contemplation the Volume of the Sacred Law [The Volume of the Sacred Law is the name Freemason's use for whatever sacred writings the individual accepts, it can mean the Bible, the Torah, the Koran, the Hindu Scripture, the Book of Mormon Etc] charging you ever to consider it as an unerring standard of truth and justice, and to regulate your actions by the divine precepts contained within it. There you will be taught the important duties you owe to God, to your neighbour, and to yourself. To God by never mentioning His name but with that awe and reverence which are due from the creature to his Creator, by imploring His aid on all your lawful undertakings and by looking to Him in every emergency for comfort and support. To your neighbour, by acting with him on the square, by rendering him every kind office which justice or mercy may require, by relieving his necessities and soothing his afflictions and by doing to him that which in similar circumstance you would wish him to do to you. And to yourself, by such prudent and well-regulated application of discipline to preserve your corporeal and mental faculties in their full energy so enabling you to exercise those talents with which God has blessed you, for the benefit of your fellow creatures and to His glory.

As a citizen of the world I am next to commend you to be exemplary in the discharge of your civil duties, by never involving yourself in any acts which may subvert the peace and good order of society; by paying due obedience to the laws of any State which may for a time become your place of residence, or afford you its protection and, above all, by never losing sight of the allegiance due to the Sovereign of your native land, ever remembering that nature has implanted in your breast sacred and indissoluble attachment towards the country from which you derived your birth and infant nurture.

This long piece of ritual, after explaining and discussing the basic Masonic principles of how a lodge is run ends with a final exhortation:

And as a last general recommendation, let me exhort you to dedicate yourself to such pursuits as may enable you to become respectable in life, useful to mankind, and an ornament to the Society of which you have just become a member. More especially that you devote time to study of such of the liberal arts and sciences as may be within the compass of your attainments, and that, without neglecting the ordinary duties of your station, you feel called upon to make a daily advancement in knowledge.

The ritual encourages a new Freemason to study and learn. Its sentiments about continual learning could have easily been written down by Bishop Spratt, indeed, in his History of the Royal Society he said of its founders:

They have studied to make it, not only an Enterprise of one season, or of some lucky opportunity; but a business of time; a steady, a lasting, an uninterrupted Work. They attempted to free it from Artifice, and humours and Passions of Sects; to render it an Instrument, whereby Mankind may obtain a Dominion over Things, and not only over one another's Judgments. And lastly they have begun to establish these reformations in Philosophy, not so much, by any solemnity of Laws, or Ostentation of Ceremonies; as by solid practice.[11]

In the Second Degree there is an even stronger reference to science. During the opening ceremony of a Fellowcraft Lodge the Master says to the assembled brethren:

Brethren before opening the Lodge in the second degree let us supplicate the Grand Geometrician of the Universe, that the rays of Heaven may shed their benign influence over us, to enlighten us in the ways of nature and science.

He then opens the lodge declaring its purpose to be:

For the improvement and instruction of Fellows of the Craft of Freemasonry.

After being instructed and tested in the secret modes of identification the new Fellowcraft receives another ritual piece of ritual encouragement. I quote part of it below:

In the former degree you had the opportunity of making yourself acquainted with the principles of moral truth and virtue, you are now permitted to extend your researches into the more hidden ways of nature and science.

Before the ceremony ends the new Fellowcraft is given another intensive piece of instruction:

The study of the liberal arts, which tends to polish and adorn the mind is strongly recommended to your attention, especially

> *the science of Geometry, which is established as the basis of our art . . . As a Fellow of the Craft, you may, in our private assemblies offer your opinions on such subjects as are regularly introduced into the lectures . . . By this privilege you may improve your intellectual powers, qualify yourself as a useful member of society and strive through researching the more hidden paths of nature and science be enabled to better know your Creator.*

There was, however, one point that had struck me when I was initiated. To pass from the First to the Second degree I had to learn a series of questions and answers, which would be put to me in Open Lodge before I was allowed to proceed. One series of responses had struck an immediate chord with me. This is the sequence:

> *Q: When were you prepared to be made a Mason?*
> *A: When the sun was at the meridian.*
> *Q: As in this country Freemason's lodges are held and candidates initiated in the evening, how do you reconcile that which at first sight appears a paradox.*
> *A: The sun being at the centre and the earth revolving around the same on its own axis and Freemasonry being diffused throughout the whole of the inhabited globe it therefore follows that the sun is always at the meridian with respect to Freemasonry.*

What had hit me so forcibly was that the lodge would not allow me to become a Fellow until I had publicly affirmed the Galilean heresy. I had to contradict the most Eminent and Reverend Lord Cardinals Inquisitors-General of the Catholic Church and their belief in the immovable throne of St Peter sitting beneath a wobbling Universe, before I would be allowed to progress in my study of nature and science. Was this really a coincidence?

I suspect not, because before I could proceed to the Master's Degree I had to learn more responses concerning the Fellow Craft Degree and within them was this confirmation:

> *Q: What are peculiar objects of research in this degree?*
> *A: The hidden mysteries of nature and science.*

At the closing of the Second Degree the new Fellow of the Craft is given a short talk which starts with these words:

> *The lecture of this degree is divided into five sections, which are devoted to the study of human science, and to tracing the goodness and majesty of the Creator by minutely analysing His works.*

Yet more confirmation that the main object of the Second Degree is to promote the study of science as a means of understanding the mind of God.

There is one last important reference to the study of science that is to be found in the early stages of the Third Degree. Before the ceremony proper started I was given a summary of the previous degrees by the Master. He used these words to sum up the Second Degree:

> *You were led in the Second Degree to contemplate the intellectual faculties and to trace them through the paths of nature and science even to the throne of God Himself. The secrets of nature and the principles of intellectual truth were then unveiled to your view.*

This antient ritual sums up the inspiration which drove Sir Robert Moray to create the Royal Society. Modern Freemasonry may be eccentric, old-fashioned and slightly out of touch but its principles are still sound. I have come to believe that it was these principles which inspired Sir Robert Moray to found the Royal Society. The scientific developments that have flowed from that act have benefited the whole world.

Now I have explained a little about Freemasonry any non-Masonic readers may be wondering what Freemason's do that is *so* secret it must be carried out behind closed doors and guarded by a man holding a drawn sword?

A few years ago if you asked a typical Freemason you would have got the answer, 'I can't tell you, it's a secret.' But very little of what goes on behind the closed and tyled (guarded by a man with a drawn sword) doors of a lodge is secret. The only secrets are the means of recognition, the passwords and tokens that enable a Freemason to identify himself at any lodge throughout the world. These are the equivalent to the PIN with which you have to identify yourself at a cash machine and nobody will learn these secrets, or how to utilise them, by reading this book.

If you do want to learn them you can always join a lodge (and that includes women, as Ladies' Freemasonry is thriving).

What I hope I have managed to do, in this appendix, is to explain some of the objectives and teaching methods which Freemasonry uses to try to improve its members. These are not secret and indeed, have been of great importance in my quest to unravel the motives behind the formation of the Royal Society.

e-Illustrations of Masonry* the web version of William Preston's *Illustrations of Masonry

The usercode is found by taking the 13th word of the Postscript, ignoring the words of the title, and adding it to the 31st word of the Appendix, also ignoring the title. Hyphenated words count as a single word for this purpose. To create your password type in both words without any spacing. For example, if the first word were 'chips' and the second were 'fish', the usercode would be 'chipsfish'. If you have followed the instructions carefully, you will find you have 12 letters in your usercode.

If you would like to access the web-book, go to http://www.robertlomas.com and click on the section entitled 'The Invisible College'. From this web-page there is a link to the web-book. When you use this link you will be asked to supply the usercode you have obtained from this book. Once you have entered the usercode, follow the instructions on the web-page to access ***e-Illustrations of Masonry***. Enjoy!

Endnotes

Prologue

1 *Weaving the Web*, Tim Berners-Lee
2 *The Royal Society 1660–1940, A History of its Administration under its Charters*, Sir Henry Lyons FRS
3 *Charles II*, Arthur Bryant
4 *Illustrated English Social History, Vol 2*, G M Trevelyan
5 *The Royal Society: Concept and Creation*, Margery Purver
6 *Illustrated English Social History: The Age of Shakespeare and the Stuart Period*, G M Trevelyan

Chapter 1

1 *Understanding the Present*, Bryan Appleyard
2 *A History of the English-Speaking Peoples, Vol II*, Sir Winston Churchill
3 *History of Western Philosophy*, Bertrand Russell
4 *The Universe of Galileo and Newton*, William Bixby
5 *The Sun in the Church*, John Heilbron
6 Ibid

7 *The Selfish Gene*, Richard Dawkins
8 *Oliver Cromwell*, John Buchan
9 *The Royal Society 1660–1940*, Sir Henry Lyons FRS
10 *The Royal Society and its Fellows 1660–1700: The Morphology of an early scientific institution*, Michael Hunter
11 *The Royal Society 1660–1940*, Sir Henry Lyons FRS

Chapter 2

1 *The Royal Society: Concept and Creation*, Margery Purver
2 *The Royal Society, Its Origins and Founders*, ed. Sir Harold Hartley
3 *The Diary of John Evelyn*, ed. E S de Beer
4 *The Mathematical and Philosophical Works of the Right Rev. John Wilkins*
5 *The Shorter Pepys*, ed. Robert Latham
6 *Science and Education in the Seventeenth Century*, Allen G Debus
7 Wallis also played a major part in founding the Royal Society
8 *Dictionary of National Biography*, ed. Sidney Lee
9 *The Royal Society and its Fellows 1660–1700: The Morphology of an early scientific institution*, Michael Hunter
10 *The Royal Society 1660–1940*, Sir Henry Lyons FRS
11 *Cromwell, Oliver*, Microsoft® Encarta® 97 Encyclopaedia. © 1993–1996 Microsoft Corporation
12 'William Viscount Brouncker, PRS' in *The Royal Society, Its Origins and Founders*, ed. Sir Harold Hartley
13 Ibid
14 *History of my own Times*, Gilbert Burnet, ed. O Airey
15 'Sir Robert Moray', D C Martin, in *The Royal Society, Its Origins and Founders*, ed. Sir Harold Hartley
16 *Short Abridgement of Britane's Distemper*, Patrick Gordon
17 *The Origins of Freemasonry*, David Stevenson
18 *History of my own Times*, Gilbert Burnet, ed. O Airey
19 *The Royal Society, Its Origins and Founders*, ed. Sir Harold Hartley
20 *Athena Oxonienses*, ed. Anthony Wood
21 *The Royal Society 1660–1640*, Sir Henry Lyons FRS
22 *The Royal Society, Its Origins and Founders*, ed. Sir Harold Hartley
23 *History of the Royal Society*, G R Weld
24 *The history of the survey of Ireland, commonly called the Down Survey by Dr W Petty*, T A Larson
25 In 1662 his sailing catamaran *Invention I* was to win a race against the packet boat, between Dublin and Holyhead
26 *The Royal Society, Its Origins and Founders*, ed. Sir Harold Hartley

27 *Brevis Assertio Systematis Saturni*, C Huygens
28 The Declaration had been dated 4 April 1660 and it promised that if restored he would, subject to the assent of parliament grant a full amnesty, liberty of conscience, settlement of all claims to landed property and pay the army the money it was owed
29 Secretary of the RS from 1752–55
30 *History of the Royal Society*, Thomas Birch
31 *Science-based Dating in Archaeology*, M J Aitken
32 'William Ball FRS', Angus Armitage in *The Royal Society, Its Origins and Founders*, ed. Sir Harold Hartley
33 *Voyages d'Angleterre*, B. de Monconys, ed. M C Henry
34 *Sir Christopher Wren: A biography*, H H Hutchison
35 'Laurence Rooke', C A Ronan in *The Royal Society, Its Origins and Founders*, ed. Sir Harold Hartley
36 *Longitude*, Dava Sobel
37 'Sir Christopher Wren, PRS', Sir John Summerson in *The Royal Society, Its Origins and Founders*, ed. Sir Harold Hartley
38 *Brief Lives*, John Aubrey, ed. A Powell
39 *Dictionary of National Biography*, ed. Sidney Lee
40 *Parentalia or Memoirs of the Family of the Wrens*, Stephen Wren
41 Ibid
42 *Lives of the Gresham Professors*, J Ward
43 'Abraham Hill FRS', R E Maddison in *The Royal Society, Its Origins and Founders*, ed. Sir Harold Hartley
44 *Titles of Patents of Invention*, ed. B Woodcroft
45 *The Royal Society 1660–1940,* Sir Henry Lyons FRS

Chapter 3

1 *The Royal Society 1660–1940*, Sir Henry Lyons FRS
2 Ibid
3 Of the twelve only Laurence Rooke was missing as he had died before a Charter was issued
4 *Biographical Encyclopaedia of Science*, I Azimov
5 *Elias Ashmole, 1617–1692*, C H Josten
6 *The Royal Society 1660–1940*, Sir Henry Lyons FRS
7 *The Origins of Freemasonry*, David Stevenson
8 *The Royal Society: Concept and Creation*, Margery Purver
9 *Robert Boyle, Works*, Thomas Birch
10 Ibid
11 Ibid

12 Ibid
13 *The Royal Society: Concept and Creation*, Margery Purver
14 *A History of the Royal Society*, Thomas Spratt
15 *The Royal Society, Its Origins and Founders*, ed. Sir Harold Hartley
16 *Treatise of Algebra*, John Wallis
17 *A Defence of the Royal Society*, John Wallis
18 *The Royal Society 1660–1940*, Sir Henry Lyons FRS
19 Sir Robert had been initiated at Newcastle in 1641

Chapter 4

1 *Francis Bacon*, Anthony Quinton
2 *A History of the Royal Society*, Thomas Spratt
3 *The Life and Works of Francis Bacon*, William Rawley
4 *A History of the Royal Society*, Thomas Spratt
5 *Francis Bacon*, Anthony Quinton
6 *Diary and Correspondence of John Evelyn FRS*, ed. William Bray
7 *The New Atlantis*, Francis Bacon (Baron Verulam)
8 *The Royal Society: Concept and Creation*, Margery Purver
9 *A History of the Royal Society*, Thomas Spratt
10 Ibid

Chapter 5

1 *A History of England 1485–1688*, Cyril E Robinson
2 *King James Bible Reproduction Edition*
3 *A History of the English-Speaking Peoples*, Winston S Churchill
4 *Gould's History of Freemasonry, Vol III*, ed. Dudley Wright
5 All with the Christian name of John
6 All the John Mylnes seemed to call their eldest sons John
7 *Gould's History of Freemasonry, Vol III*, ed. Dudley Wright
8 *Genealogie of the Saintclaires of Rosslyn, including the Chartulary of Rosslyn*, R A Hey
9 Ibid
10 *The Second Messiah*, C Knight and R Lomas
11 *The Origins of Freemasonry*, David Stevenson
12 *William Elphinstone, his college chapel and the second of April*, G P Edwards, Aberdeen University Review, 1i (1985) p1–17
13 *The Origins of Freemasonry*, David Stevenson
14 Ibid

15 *Accounts of the Masters of Works*, ed. H M Paton, i, xvii; 1581–4, no 1676, Edinburgh, 1957
16 *Accounts of the Masters of Works*, ed. H M Paton, 1556–67, Edinburgh, 1957
17 *The Origins of Freemasonry*, David Stevenson
18 *Calendar of State Papers relating to Scotland*, 1581–3, 13 Vols, HMSO, London, 1910–59
19 *Calendar of State Papers relating to Scotland*, 1584–5, 13 Vols, HMSO, London, 1910–59
20 Scottish Record Office, PS. I/61, Register of the privy seal
21 *Historical Sketch of the Grand Lodge of Antient Free and Accepted Masons of Scotland*, Appendix 1, Grand Lodge of Antient Free and Accepted Masons of Scotland, 1986
22 Ibid
23 Ibid
24 Modern Lodges place great emphasis on written minutes to prove their seniority as the opening quotation, from Masonic historian Dudley Wright, shows
25 What Masons refer to as a cowan
26 Later he became Earl of Dunfermline, and was a member of Aberdeen Lodge
27 *Memorie of the Somervilles; being a history of the baronial house of Somerville*, Lord John Somerville, ed. Sir Walter Scott
28 *The Origins of Freemasonry*, David Stevenson
29 Later Stuart kings would create a Masonic Order in Scotland which automatically makes the King of Scots its Grand Master
30 *The Origins of Freemasonry*, David Stevenson
31 Where the oldest evidence of Masonic ritual is carved into the wall
32 *The Worship of the Reformed Church of Scotland 1550–1638*, W McMillan, London, 1931
33 *The Records of the Synod of Lothian and Tweeddale 1589–96*, ed. J Kirk, Stair Society, Edinburgh, 1977
34 Ibid
35 *Genealogie of the Saintclaires of Rosslyn, including the Chartulary of Rosslyn*, R A Hey, J Maidment, Edinburgh, 1835
36 There seem to be a lot of William Sinclairs in this story!
37 *Historical Sketch of the Grand Lodge of Antient Free and Accepted Masons of Scotland*, Grand Lodge of Antient Free and Accepted Masons of Scotland, 1986
38 Clockmaker to the king and gentleman of the Privy Chamber

39 *Gould's History of Freemasonry*, ed. Dudley Wright
40 *Charles I*, Christopher Hibbert
41 *A History of the English-Speaking Peoples Vol II*, Winston S Churchill
42 *The Stuarts*, J P Kenyon
43 *The Early Stuarts*, Godfrey Davies
44 *A History of England, 1485–1688*, Cyril E Robinson
45 *Illustrations of Masonry*, William Preston

Chapter 6

1 *The Origins of Freemasonry*, David Stevenson
2 *The Life of Sir Robert Moray, Soldier, Statesman and Man of Science*, A Robinson
3 *History of England*, Leopold von Ranke
4 *A History of the English-Speaking Peoples*, Winston S Churchill
5 Ibid
6 Ibid
7 *Charles I*, Christopher Hibbert
8 *Dictionary of National Biography*, ed. Sidney Lee
9 Kincardine Letters, f 48r
10 *Travels in Scotland*, Hume Brown
11 *The Origins of Freemasonry*, David Stevenson
12 *Masonry, symbolism and ethics in the life of Sir Robert Moray, FRS*, David Stevenson, PSAS, 114 (1984) p401–431
13 Kincardine Letters, f 31v
14 *The Royal Society, its Origins and Founders*, ed. Sir Harold Hartley
15 Ibid
16 *A Short Abridgement of Britane's Distemper*, Patrick Gordon
17 *The Royal Society 1660–1940*, Sir Henry Lyons FRS
18 *The History of the Rebellion and Civil Wars in England, by Edward, Earl of Clarendon*, ed. W Dunn Macray
19 *A Short Abridgement of Britane's Distemper*, Patrick Gordon
20 *Dictionary of National Biography*, ed. Sidney Lee
21 The minute of the meeting, which is kept at the Grand Lodge of Scotland, is reproduced as photofacsimile in Gould's *History of Freemasonry*
22 *Dictionary of National Biography*, ed. Sidney Lee
23 *Gould's History of Freemasonry Vol III*, ed. Dudley Wright
24 Kincardine Letters, f 36v
25 *A History of the English-Speaking Peoples*, Winston S Churchill
26 *Dictionary of National Biography*, ed. Sidney Lee

27 *Charles II*, Osmund Airey
28 *The Origins of Freemasonry*, David Stevenson
29 Ibid
30 *Charles I*, Christopher Hibbert
31 *The Tragedy of Charles II*, Hester W Chapman
32 *The Origins of Freemasonry*, David Stevenson
33 *The Royal Society, its Origins and Founders,* ed. Sir Harold Hartley
34 *The Tragedy of Charles II*, Hester W Chapman
35 *Charles II*, Osmund Airey
36 This was the deal that was used by Cromwell as evidence of Treason and Charles lost his head over it
37 *Mercurius Britanicus*, a Parliamentary Broadsheet of the time
38 *Hamilton Papers*, ed. S R Gardiner, Camden Society, London, 1880
39 *King Charles II*, Antonia Fraser
40 *The Tragedy of Charles II*, Hester W Chapman
41 *Letters, 1623–1673, Henrietta Maria*, ed. Mary A Everret
42 *The Royal Society, its Origins and Founders,* ed. Sir Harold Hartley
43 *Dictionary of National Biography*, ed. Sidney Lee
44 *The Origins of Freemasonry*, David Stevenson
45 *Gould's History of Freemasonry Vol II*, ed. Dudley Wright
46 *The Tragedy of Charles II*, Hester W Chapman
47 *Autobiography*, A Moray (Murray)
48 *Memoires,* Mme de Montpensier, ed. A Cheruel
49 *Celtic Gods, Celtic Goddesses,* R J Stewart
50 *Symbols of our Celtic-Saxon Heritage,* W J Bennett
51 The Calendar of Domestic State Papers, 28 August 1650, Scottish National Records Office, Edinburgh
52 *Personal History of King Charles II 1650–1651*, Charles J Lyon
53 Ibid
54 *Dictionary of National Biography*, ed. Sidney Lee
55 Moray would propose him for Fellowship of the Royal Society within a week of its first meeting
56 *The Origins of Freemasonry*, David Stevenson
57 *The Great Marquess*, John Willcock
58 Or Murray as it is sometimes spelt
59 *Personal History of King Charles II 1650–1651*, Charles J Lyon
60 *Moreton Muniments*, Scottish Records Office, GD 150/2949
61 *Personal History of King Charles II 1650–1651*, Charles J Lyon
62 National Library of Scotland, Adv, MS 29.2.9, Balcarres papers, vol ix, ff 243–2

63 *The Origins of Freemasonry*, David Stevenson
64 *King Charles II*, Antonia Fraser
65 *The Royal Society, its Origins and Founders*, ed. Sir Harold Hartley
66 *History of My Own Times*, G Burnet, ed. O Airey
67 Ibid

Chapter 7

1 *Elias Ashmole (1617–1692) His Autobiographical and Historical Notes*, ed. C H Josten
2 Ibid
3 *The Royal Society and its Fellows 1660–1700*, Michael Hunter
4 Ibid
5 Royal Society, Council Minutes Copy, Vol I, 1673–82, p238
6 The Diary of Robert Hooke, 7 March 1676, MS Guildhall Library
7 I have already mentioned that the first documented Freemasonic initiation on English soil happened in 1641; and that Freemason was Sir Robert Moray
8 *Elias Ashmole (1617–1692) His Autobiographical and Historical Notes*, ed. C H Josten
9 Ibid
10 Ibid
11 The University Register contains no records of Brasenose graduations between 26 June 1643 and 13 November 1646
12 *The Brazen Nose*, R W Jeffery
13 MS Ashmole, 1136, f 16
14 MS Ashmole, 313, art I
15 *Gould's History of Freemasonry Vol II*, ed. Dudley Wright
16 *Early History and Antiquities of Freemasonry*, W H Rylands
17 *Dictionary of National Biography*, ed. Sidney Lee
18 *Elias Ashmole (1617–1692) Biographical Introduction*, C H Josten
19 *Elias Ashmole (1617–1692) Vol II*, C H Josten
20 *Dictionary of National Biography*, ed. Sidney Lee
21 Seth Ward was the man who had given John Wilkins a London home when he was thrown out of Cambridge. He was also the tutor of Viscount William Brouncker, the first President of the Royal Society
22 *A Defence of the Royal Society*, John Wallis
23 F313 RS proposed by Viscount Brouncker
24 *Elias Ashmole (1617–1692) Biographical Introduction*, C H Josten
25 *Elias Ashmole (1617–1692) Vol IV*, C H Josten
26 F70, RS

27 *The Royal Society and its Fellows 1660–1700*, Michael Hunter
28 *The Royal Society, Its Origins and Founders*, ed. Sir Harold Hartley
29 *Gresham College: Precursor of the Royal Society,* F R Johnson, Journal of the History of Ideas, Vol I pp 423–38, New York, 1940
30 *The Royal Society, Its Origins and Founders*, ed. Sir Harold Hartley
31 *The Origins of the Three Degrees of Craft Freemasonry*, V R M Langton, Year Book of the Grand Lodge of Antient Free and Accepted Masons of Scotland, Edinburgh, 1995
32 *Illustrations of Masonry*, William Preston
33 Ibid
34 Ibid
35 Ibid
36 Ibid
37 Ibid
38 *Elias Ashmole 1617–1692*, ed. C H Josten
39 *Charles I*, Christopher Hibbert
40 *Manual of Freemasonry*, Anon

Chapter 8

1 *Oliver Cromwell*, John Buchan
2 *The Shorter Pepys*, ed. Robert Latham
3 *King Charles II*, Arthur Bryant
4 *The Tragedy of Charles II*, Hester W Chapman
5 *Dictionary of National Biography*, ed. Sidney Lee
6 *The Shorter Pepys*, ed. Robert Latham
7 *King Charles II*, Arthur Bryant
8 *The Diary of John Evelyn*, ed. E S de Beer
9 *The Letters, Speeches and Proclamations of King Charles I*, Sir Charles Petrie
10 *The Earl of Manchester's Speech to His Majesty in the name of the Peers. At his arrival at Whitehall 29th May 1660, With his Majesties Gracious Answer.* Printed by John Macock and Francis Tyton, 1660
11 *The Tragedy of Charles II*, Hester W Chapman
12 Ibid
13 *Memoirs of the Life of Colonel Hutchinson*, Lucy Hutchinson
14 *The True Historical Narrative of the Rebellion and Civil Wars in England*, Edward Hyde, ed. W Dunn Macray
15 *Dictionary of National Biography*, ed. Sidney Lee
16 *The Tragedy of Charles II*, Hester W Chapman
17 *Dictionary of National Biography*, ed. Sidney Lee
18 *The Tragedy of Charles II*, Hester W Chapman

19 *King Charles II*, Antonia Fraser
20 'The Savoy Conference and the Revision of the Book of Common Prayer', E C Rafcliff, in *From Uniformity to Unity*, ed. G F Nuttal
21 *Calendar of the State Papers Domestic, 1660,* Official Records Office Kew
22 *The Tragedy of Charles II*, Hester W Chapman
23 *A History of the English-Speaking Peoples Vol II*, Winston S Churchill
24 *Orange and Stuart 1641–72*, Pieter Geyl
25 Kincardine Papers, ff 160v, 190r, 194r also quoted in Stevenson
26 *Imperial Commonwealth*, Lord Elton
27 Ibid
28 Ibid
29 *The Royal Society: Concept and Creation*, Margery Purver
30 *The Origins of Freemasonry*, David Stevenson
31 *History of My Own Times*, G Burnet, ed. O Airey
32 *Ingenious Pursuits,* Lisa Jardine
33 *The Royal Society, Its Origins and Founders,* ed. Sir Harold Hartley

Chapter 9

1 *The Royal Society 1660–1940*, Sir Henry Lyons FRS
2 *The Diary of John Evelyn*, ed. E S de Beer
3 Journal Book of the Royal Society, 5 December 1660
4 *The Shorter Pepys,* ed. R Latham
5 Ibid
6 *The Royal Society 1660–1940*, Sir Henry Lyons FRS
7 Journal Book of the Royal Society, 19 December 1660
8 Ibid
9 Journal Book of the Royal Society, 16 January 1661
10 *A History of the Royal Society*, Thomas Spratt
11 Journal Book of the Royal Society, 20 March 1661
12 *The Royal Society: Concept and Creation*, Margery Purver
13 *The Royal Society 1660–1940,* Sir Henry Lyons FRS
14 *The Second Creation*, Ian Wilmut
15 *The Royal Society, Its Origins and Founders*, ed. Sir Harold Hartley
16 This would eventually lead to Evelyn publishing the very first book on forestry
17 *Ingenious Pursuits*, Lisa Jardine
18 *An Examen of Mr T Hobbes his Dialogue Physicus De Nature Aeris*, Robert Boyle
19 *The Royal Society 1660–1940*, Sir Henry Lyons FRS
20 *Profit and Power, A study of England and the Dutch Wars*, Charles Wilson

21 *King Charles II*, Antonia Fraser
22 *The King my Brother*, C H Hartman
23 *The Rise of the English Shipping Industry in the 17th and 18th Century*, Ralph Davies
24 *The Diary of John Evelyn*, ed. E S de Beer
25 *The Royal Society: Concept and Creation*, Margery Purver
26 *Charta Secunda*, 22 April 1663, Royal Society
27 *History of the Royal Society*, G R Weld
28 *Charta Secunda*, 22 April 1663, Royal Society
29 Royal Society Journal Book 1
30 *Oeuvres Completes*, Christiaan Huygens, The Hage 1889, Vol IV p343
31 *The Royal Society and its Fellows 1660–1700*, Michael Hunter
32 *The Royal Society 1660–1940*, Sir Henry Lyons FRS
33 *Charta Secunda*, 22 April 1663, Royal Society
34 *The Royal Society: Concept and Creation*, Margery Purver

Chapter 10

1 *The Royal Society and its Fellows 1660–1700*, Michael Hunter
2 *Illustrations of Masonry*, William Preston
3 *The Shorter Pepys*, ed. Robert Latham
4 *King Charles II*, Antonia Fraser
5 British Library, Add Ms 15858
6 Newcastle, 1641
7 Kincardine Papers f 58r
8 *The Origins of Freemasonry*, David Stevenson
9 *The Lauderdale Papers*, ed. O Airey
10 Ibid
11 *The Origins of Freemasonry*, David Stevenson
12 Ritual of the Degree of the Royal Arch of Enoch
13 *The Royal Society 1660–1940*, Sir Henry Lyons FRS
14 *Ingenious Pursuits*, Lisa Jardine
15 Ibid
16 *The Royal Society, Its Origins and Founders*, ed. Sir Harold Hartley
17 *Ingenious Pursuits*, Lisa Jardine
18 *The Royal Society and its Fellows 1660–1700*, Michael Hunter
19 *A History of the English-Speaking Peoples Vol II*, Winston S Churchill
20 *The Royal Society, Its Origins and Founders*, ed. Sir Harold Hartley
21 *The Life of Sir William Petty 1623–1687*, Edmond Fitzmaurice
22 *King Charles II*, Antonia Fraser
23 *The Royal Society 1660–1940*, Sir Henry Lyons FRS

24 Ibid
25 *The Royal Society, Its Origins and Founders*, ed. Sir Harold Hartley
26 *The Origins of Modern Science*, Herbert Butterfield
27 *The Royal Society, Its Origins and Founders*, ed. Sir Harold Hartley
28 *The Works of the Honourable Robert Boyle*, Robert Boyle
29 *The Royal Society, Its Origins and Founders*, ed. Sir Harold Hartley
30 *The Origins of Modern Science*, Herbert Butterfield
31 *The Royal Society, Its Origins and Founders*, ed. Sir Harold Hartley

Chapter 11

1 *The Shorter Pepys*, ed. Robert Latham
2 Ibid
3 *The arrest and imprisonment of Henry Oldenburg,* D Mackie, Notes and Records, 6, p28 (1949)
4 *The Royal Society and its Fellows 1660–1700*, Michael Hunter
5 Ibid
6 *The Royal Society, Its Origins and Founders*, ed. Sir Harold Hartley
7 *Illustrations of Masonry*, William Preston
8 *The Passionate Shepherdess: Aphra Behn 1640–1689*, Maureen Duffy
9 *King Charles II*, Antonia Fraser
10 *The arrest and imprisonment of Henry Oldenburg,* D Mackie, Notes and Records, 6, p28 (1949)
11 *Imperial Commonwealth*, Lord Elton
12 *The Origins of Modern Science*, Herbert Butterfield
13 *Life and Times of the Duchess of Portsmouth*, Jeanine Delpech
14 *The Court Wits of the Restoration*, J H Wilson
15 *King Charles II*, Antonia Fraser
16 *Greenwich Millennium*, Clive Aslet
17 *The Royal Society 1660–1940*, Sir Henry Lyons FRS

Chapter 12

1 *All Scientists Now,* Marie Boas Hall
2 'King Charles II', E S de Beer, in *The Royal Society, Its Origins and Founders*, ed. Sir Harold Hartley
3 Ibid
4 *The Royal Society, Its Origins and Founders*, ed. Sir Harold Hartley
5 *A History of the English-Speaking Peoples Vol II*, Winston S Churchill
6 Anne had been the second daughter of James II by his first wife, the Protestant daughter of the Earl of Clarendon

7 *A History of the English-Speaking Peoples Vol II*, Winston S Churchill
8 Ibid
9 UGLE Masonic Year Book 1994–95
10 Minutes of the Lodge of Edinburgh (St Mary's) (Metropolitian) No 1 March 23, 1684
11 Records of the Grand Lodge of Antient Free and Accepted Masons of Scotland
12 *Gould's History of Freemasonry*, ed. Dudley Wright
13 *Our Separated Brethren The Freemasons*, Alex Mellor
14 *Gould's History of Freemasonry*, ed. Dudley Wright
15 *Our Separated Brethren The Freemasons*, Alex Mellor
16 *A History of the English-Speaking Peoples Vol II*, Winston S Churchill
17 Mackey: Encyclopaedia of Freemasonry
18 Hansard 1799
19 *Gould's History of Freemasonry*, ed. Dudley Wright
20 Ibid
21 Ibid
22 'Sir Robert Moray', D C Martin, in *The Royal Society, Its Origins and Founders*, ed. Sir Harold Hartley
23 *The Origins of Freemasonry*, David Stevenson
24 *Elias Ashmole, 1617–1692*, ed. C H Josten
25 *All Scientists Now*, Marie Boas Hall
26 Ibid
27 Ibid
28 Ibid
29 Ibid
30 Ibid
31 *The Royal Society 1660–1940*, Sir Henry Lyons FRS
32 *All Scientists Now*, Marie Boas Hall
33 Ibid
34 Ibid

Chapter 13

1 *Our Separated Brethren The Freemasons*, Alex Mellor
2 *A History of the Royal Society*, Thomas Spratt
3 *Book of Constitutions*, James Anderson
4 Scottish Records Office, GD.406.1.C.6132
5 *The Origins of Freemasonry*, David Stevenson
6 Ibid
7 *Scientists and Amateurs: A history of the Royal Society*, D Stimson

Postscript

1 *A Shorter Intermediate Mechanics*, D Humphrey & J Topping
2 *Mathematical Principles of Natural Philosophy*, Sir Isaac Newton
3 Ibid
4 Ibid
5 *The Large Scale Structure of Space Time,* S Hawking & G Ellis
6 *A Brief History of Time*, S Hawking
7 *Mathematical Principles of Natural Philosophy*, Sir Isaac Newton

Appendix

1 Taken from the leaflet issued by the Provincial Grand Lodge of Aberdeen City and distributed freely to all visitors to the Aberdeen Masonic Temple
2 *The Origins of the Three Degrees of Craft Freemasonry*, V R M Langton, Year Book of the Grand Lodge of Antient Free and Accepted Masons of Scotland, Edinburgh, 1995
3 Ibid
4 *Ars Quatuor Coronatorum*, A M Mackay, Vol XXV, p278, 1912
5 *Gould's History of Freemasonry*, ed. Dudley Wright
6 *The Origins of the Three Degrees of Craft Freemasonry*, V R M Langton, Year Book of the Grand Lodge of Antient Free and Accepted Masons of Scotland, Edinburgh, 1995
7 *Gould's History of Freemasonry Vol II,* Dudley Wright
8 *The Secret Traditions of Freemasonry*, A E Waite
9 *The Origins of Freemasonry*, David Stevenson
10 Ibid
11 *A History of the Royal Society*, Thomas Spratt

Bibliography

Airey, Osmund (ed.), *The Lauderdale Papers*, Camden Society, London, 1884

Airey, Osmund, *Charles II*, Cassell & Co, London, 1901

Aitken, M J, *Science-based Dating in Archaeology*, Longman, London, 1990

Anderson, James, *Book of Constitutions,* Grand Lodge of London, London, 1723

Anon, *Manual of Freemasonry*, London, 1732

Anon, *The Scottish Ritual of Craft Freemasonry*, J Bethune, Edinburgh, 1906

Appleyard, Bryan, *Understanding the Present*, Pan, London, 1992

Armstrong, John, *History of the Lodge of Lights No 148 Warrington with a Preliminary Chapter on Early Masonry and the Initiation of Elias Ashmole in Warrington*, Lodge of Lights, Warrington, 1898

Ashley, Maurice, *England in the Seventeenth Century*, Penguin, London, 1952

Aslet, Clive, *Greenwich Millennium*, Fourth Estate, London, 2000
Atkinson, Dwight, *Scientific Discourse in Sociohistorical Context: the Philosophical Transactions of the Royal Society of London*, Lawrence Erlbaum, New Jersey, 1999
Aubrey, John, A Powell (ed.), *Brief Lives*, Cape, London, 1949
Bacon, Francis (Baron Verulam), *The New Atlantis*, London, 1627
Bennett, W J, *Symbols of our Celto-Saxon Heritage*, Covenant Books, Edinburgh 1976
Berners-Lee, Tim, *Weaving the Web*, Orion Business Books, London, 1999
Birch, Thomas, *History of the Royal Society*, The Royal Society, London, 1744
Birch, Thomas, *Robert Boyle, Works*, The Royal Society, London, 1744
Bixby, William, *The Universe of Galileo and Newton*, Cassell, London, 1964
Boas Hall, Marie, *All Scientists Now*, Cambridge University Press, Cambridge, 1984
Boyle, Robert, *An Examen of Mr T Hobbes his Dialogue Physicus De Nature Aeris*, London, 1662
Boyle, Robert, *The Works of the Honourable Robert Boyle*, London, 1744
Bray, William (ed.), *Diary and Correspondence of John Evelyn, FRS*, London, 1850
Brown, Hume, *Travels in Scotland*, Edinburgh, 1618
Bryant, Arthur, *Charles II*, Collins, London, 1960
Buchan, John, *Oliver Cromwell*, The Reprint Society, London, 1941
Burnet, Gilbert, *History of my own Times*, London, 1723
Butterfield, Herbert, *The Origins of Modern Science*, Bell & Hyman, London, 1957
Chapman, Hester W, *The Tragedy of Charles II*, Jonathan Cape, London, 1964
Cheruel, A (ed.), *Memoires*, Mme de Montpensier, Paris, 1892
Churchill, Sir Winston, *A History of the English-Speaking Peoples, Vol II*, Cassell, London, 1954
da Costa Andrade, Edward Neville, *A Brief History of the Royal Society*, The Royal Society, London, 1960
Davies, Godfrey, *The Early Stuarts*, Oxford University Press, Oxford, 1968

Davies, Ralph, *The Rise of the English Shipping Industry in the 17th and 18th Century*, Clarendon Press, London, 1962

Dawkins, Richard, *The Selfish Gene*, Oxford University Press, Oxford, 1976

de Beer, E S (ed.), *The Diary of John Evelyn*, Oxford University Press, Oxford, 1959

Debus, Allen G, *Science and Education in the Seventeenth Century*, Macdonald, London, 1970

Delpech, Jeanine, *Life and Times of the Duchess of Portsmouth*, Collins, London, 1953

Duffy, Maureen, *The Passionate Shepherdess: Aphra Behn 1640–1689*, Cambridge University Press, London, 1977

Dunn Macray, W (ed.), *The History of the Rebellion and Civil Wars in England, by Edward, Earl of Clarendon*, Oxford University Press, Oxford, 1827

Edwards, G P, *William Elphinstone, his college chapel and the second of April*, Aberdeen University Review, 1i (1985) p1–17

Elton, Lord, *Imperial Commonwealth*, Collins, London, 1945

Everret, Mary A (ed.), *Letters, 1623–1673, Henrietta Maria*, Oxford University Press, Oxford, 1958

Fitzmaurice, Edmond, *The Life of Sir William Petty 1623–1687*, London, 1897

Fraser, Antonia, *King Charles II*, Weidenfeld and Nicolson, London, 1979

Gardiner, S R (ed.), *Hamilton Papers*, Camden Society, London, 1880

Geyl, Pieter, *Orange and Stuart 1641–72*, Weidenfield and Nicolson, London 1969

Gleason, Mary Louise, *The Royal Society of London; Years of Reform 1827–1847*, Garland, London, 1991

Gordon, Patrick, *A Short Abridgement of Britane's Distemper*, Spalding Club, London, 1844

Hartley, Sir Harold (ed.), *The Royal Society, Its Orgins and Founders*, The Royal Society, London, 1960

Hartman, C H, *The King my Brother*, Macdonald, London, 1954

Hawking, S & Ellis, G, *The Large Scale Structure of Space Time*, Cambridge University Press, Cambridge, 1973

Hawking, S, *A Brief History of Time*, Bantam, London, 1988

Heilbron, John, *The Sun in the Church*, Harvard University Press, New York, 2000

Hey, R A, *Genealogie of the Saintclaires of Rosslyn, including the Chartulary of Rosslyn*, J Maidment, Edinburgh, 1835

Hibbert, Christopher, *Charles I*, Weidenfield and Nicolson, London, 1968

Hooke, Robert, *The Diary of Robert Hooke*, MS Guildhall Library

Humphrey, D & Topping, J, *A Shorter Intermediate Mechanics*, Longmans, London, 1962

Hunter, Michael, *Establishing the New Science: the experience of the early Royal Society*, Boydell Press, Woodridge, 1989

Hunter, Michael, *The Royal Society and its Fellows 1660–1700: The Morphology of an early scientific institution*, The British Society for the History of Science, London, 1982

Hutchinson, Lucy, *Memoirs of the Life of Colonel Hutchinson*, London, 1806

Hutchison, H H, *Sir Christopher Wren: A biography*, Victor Gollancz, London, 1976

Huygens, C, *Brevis Assertio Systematis Saturni*, The Hague, 1662

Huygens, Christiaan, *Oeuvres Completes*, The Hague, 1889

Hyde, Edward, Dunn Macray, W (ed.), *The True Historical Narrative of the Rebellion and Civil Wars in England*, Oxford University Press, Oxford, 1888

Jardine, Lisa, *Ingenious Pursuits*, Abacus, London, 1999

Jeffery, R W, *The Brazen Nose*, Oxford University Press, Oxford, 1927

Johnson, F R, *Gresham College: Precursor of the Royal Society*, Journal of the History of Ideas, Vol I pp 423–38, New York, 1940

Josten, C H, *Elias Ashmole (1617–1692) His Autobiographical and Historical Notes*, Oxford University Press, Oxford, 1966

Kenyon, J P, *The Stuarts*, B T Batsford, London, 1958

Kirk, J (ed.), *The Records of the Synod of Lothian and Tweeddale 1589–96*, Stair Society, Edinburgh, 1977

Lancaster-Brown, Peter, *Halley and His Comet*, Blandford Press, Poole, 1985

Langton, V R M, *The Origins of the Three Degrees of Craft Freemasonry*, Year Book of the Grand Lodge of Antient Free and Accepted Masons of Scotland, Edinburgh, 1995

Larson, T A, *The history of the survey of Ireland, commonly called the Down Survey by Dr W Petty*, Irish Archaeological Society, Dublin, 1851

Latham, Robert (ed.), *The Shorter Pepys*, Bell & Hyman, London, 1985

Lee, Sidney (ed.), *Dictionary of National Biography*, Smith, Elder & Co, London, 1903

Linklater, Eric, *The Royal House of Scotland*, Sphere Books, London, 1970

Lyon, Charles J, *Personal History of King Charles II 1650–1651*, Edinburgh, 1851

Lyons FRS, Sir Henry, *The Royal Society 1660–1940*, Cambridge University Press, Cambridge, 1944

Lyons FRS, Sir Henry, *The Royal Society 1660–1940, A History of its Administration under its Charters*, Greenwood Press, New York, 1968

Lyons FRS, Sir Henry (ed.), *The Royal Society, The record of the Royal Society of London for the promotion of Natural Knowledge*, The Royal Society, London, 1940

Mackay, A M, *Ars Quatuor Coronatorum*, Vol XXV, p278, 1912

Mackie, D, *The arrest and imprisonment of Henry Oldenburg*, Notes and Records, 6, p28, 1949

Matheson, P E & E F (eds), *Francis Bacon – Selections*, Clarendon Press, Oxford, 1922

Mellor, Alex, *Our Separated Brethren The Freemasons*, George G Harrop, London, 1964

de Monconys, B, Henry, M C (ed.), *Voyages d'Angleterre*, Paris, 1887

Moray (Murray), A, *Autobiography*, Cassell & Co, London, 1875

Newton, Sir Isaac, *Mathematical Principles of Natural Philosophy*, Great Books of the World, Encylopaedia Britannica, Inc. London, 1952

Nuttal, G F, (ed.), *From Uniformity to Unity*, Oxford University Press, Oxford, 1962

Paton, H M (ed.), *Accounts of the Masters of Works*, i, xvii; 1581–4, no 1676, Edinburgh, 1957

Petrie, Sir Charles, *The Letters, Speeches and Proclamations of King Charles I*, Cassell, London, 1935

Popper, Karl R, *The Logic of Scientific Discovery*, Hutchinson, London, 1959

Preston, William, *Illustrations of Masonry*, T Wilkie, London, 1772

Purver, Margery, *The Royal Society: Concept and Creation*, Routledge and Kegan Paul, London, 1967

Quinton, Anthony, *Francis Bacon*, Oxford University Press, Oxford, 1980

Rawley, William, *The Life and Works of Francis Bacon*, Cambridge University, Cambridge, 1621

Robinson, A, *The Life of Sir Robert Moray, Soldier, Statesman and Man of Science*, Caxton & Co, London, 1922

Robinson, Cyril E, *A History of England 1485–1688*, Methuen, London, 1925

Russell, Bertrand, *History of Western Philosophy*, Unwin University Books, London, 1946

Russell, Bertrand, *The Impact of Science on Society*, Unwin, London, 1952

Rylands, W H, *Early History and Antiquities of Freemasonry*, Caxton, London 1885

Sobel, Dava, *Longitude*, Fourth Estate, London, 1995

Somerville, Lord John, Scott, Sir Walter (ed.), *Memorie of the Somervilles; being a history of the baronial house of Somerville*, Edinburgh, 1815

Sprat, Thomas, *A History of the Royal Society*, The Royal Society, London, 1667

Stevenson, David, *Masonry, symbolism and ethics in the life of Sir Robert Moray, FRS*, PSAS, 114 (1984) p401–431

Stevenson, David, *The Origins of Freemasonry*, Cambridge University Press, Cambridge, 1988

Stewart, R J, *Celtic Gods, Celtic Goddesses*, Blandford, London, 1990

Stimson, D, *Scientists and Amateurs: A history of the Royal Society*, Royal Society, London, 1949

Thompson, William P L, *History of Orkney*, The Mercat Press, Edinburgh 1987

Thomson, Thomas, *History of the Royal Society*, London, 1812

Trevelyan, G M, *England Under the Stuarts*, Pelican, London, 1960

Trevelyan, G M, *Illustrated English Social History: The Age of Shakespeare and the Stuart Period*, Longmans, London, 1942

Trevelyan, G M, *Illustrated English Social History, Vol 2*, Longmans, London, 1944

Valle, Ellen, *A Collective Intelligence: the life sciences in the Royal Society as a scientific discourse community, 1665–1965*, University of Turku, Turku, 1999

von Ranke, Leopold, *History of England*, Sands & Co, London, 1875

Waite, A E, *The Secret Traditions of Freemasonry*, Caxton & Co, London, 1910

Wallis, John, *A Defense of the Royal Society*, London, 1678

Ward, J, *Lives of the Gresham Professors*, London, 1740

Wedgwood, C V, *The Trial of Charles I*, Collins, London, 1964

Weld, G R, *History of the Royal Society*, The Royal Society, London, 1848

Willcock, John, *The Great Marquess*, Cassell & Co, London, 1903

Wilmut, Ian, Campbell, Keith and Tudge, Colin, *The Second Creation*, Headline, 2000

Wilson, Charles, *Profit and Power, A study of England and the Dutch Wars*, Oxford University Press, Oxford, 1957

Wilson, J H, *The Court Wits of the Restoration*, Princeton University Press, Princeton, 1948

Wood, Anthony (ed.), *Athena Oxonienses*, Oxford, 1670

Woodcroft, B (ed.), *Titles of Patents of Invention*, Public Record Office, London, 1854

Wren, Stephen, *Parentalia or Memoirs of the Family of the Wrens*, London, 1750

Wright, Dudley (ed.), *Gould's History of Freemasonry, Vol III*, Caxton, London, 1900

Other Publications

The Calendar of Domestic State Papers, 28 August, 1650, Scottish National Records Office, Edinburgh

Calendar of the State Papers Domestic, 1660, Official Records Office, Kew

Calendar of State Papers relating to Scotland, 1581–3, 13 Vols, HMSO, London, 1910–59

Charta Secunda, 22 April 1663, Royal Society

The Earl of Manchester's Speech to His Majesty in the name of the Peers. At his arrival at Whitehall 29th May 1660, With his Majesties Gracious

Answer. Printed by John Macock and Francis Tyton, 1660
Historical Sketch of the Grand Lodge of Antient Free and Accepted Masons of Scotland, Grand Lodge of Antient Free and Accepted Masons of Scotland, 1986
Journal Book of the Royal Society, 5 December 1660
The Mathematical and Philosophical works of the Right Rev. John Wilkins, London, 1708
Minutes of the Lodge of Edinburgh (St Mary's) (Metropolitian) No 1 23 March, 1684
Moreton Muniments, Scottish Records Office, GD 150/2949
Royal Society, Council Minutes Copy, Vol I, 1673–82

Timeline

Year	*Event*
1558	Elizabeth I becomes Queen of England
1561	Francis Bacon born
1566	James VI born
1573	Francis Bacon goes to Trinity College, Cambridge
1576	Mary abdicates and James VI becomes King of Scotland
1579	Francis Bacon enrols at Grays Inn
1581	James VI starts to rule Scotland; James VI signs the Covenant
1582	Francis Bacon becomes a Barrister; James VI kidnapped by Earl of Gowrie

Year	*Event*
1584	Francis Bacon enters Parliament as Member for Melcombe Regis
1586	James VI signs Treaty of Berwick with Elizabeth I
1588	Spanish Armada defeat leaves England ruling the seas
1589	James VI marries Anne of Denmark
1590	James VI confirms Patrick Copland, Regional Warden of Masonry in Aberdeen
1596	Gresham College founded by Sir Thomas Gresham
1597	Francis Bacon publishes first essays
1598	First Schaw Statutes Issued; Francis Bacon arrested for Debt; Lectures start at Gresham College
1599	Second Schaw Statutes Issued
1600	Earl of Essex tried for Contempt; Francis Bacon secures his Freedom; Charles I born in Dunfermline; William Gilbert publishes *The Great Magnet of the Earth*
1601	Earl of Essex beheaded for Treason; James VI made a Mason at Lodge of Perth and Scoon; First Sinclair Charter
1602	William Schaw Dies
1603	Elizabeth I dies; James VI of Scotland becomes King of England
1605	Gunpowder Plot; Francis Bacon publishes *The Advancement of Learning*
1606	Francis Bacon marries Alice Barnham
1609	Robert Moray born
1610	Francis Bacon writes *The New Atlantis*

Year	*Event*
1611	Francis Bacon acts as mediator between James VI and Parliament
1613	Francis Bacon appointed Attorney General
1618	Francis Bacon made Lord Chancellor; Francis Bacon made Lord Verulam
1620	Francis Bacon publishes *Novum Organum, Indications respecting the Interpretation of Nature*
1621	Francis Bacon made Viscount St Albans; Francis Bacon tried for Bribery and found guilty; Francis Bacon pardoned for corruption
1622	Francis Bacon publishes *Historia Ventorum*
1623	Francis Bacon publishes *Historia Vitae et Mortis*
1625	James VI dies; Charles I becomes King of England
1626	Francis Bacon dies aged 65
1627	Robert Moray first records his interest in science
1628	Second Sinclair Charter Issued; Parliament Issues Petition of Right
1629	Charles I dismisses Parliament
1630	Charles II born
1633	Robert Moray joins French Scots Guard
1637	Charles I attempts to impose Anglican liturgy on Scots
1638	The Covenant of 1581 revived in Scotland
1641	Robert Moray made a Mason at Newcastle; Charles II takes seat in House of Lords as Prince of Wales; Robert Moray recruits Scots soldiers to serve in France
1642	Civil War starts with Battle of Edgehill

Year	*Event*
1643	Robert Moray knighted by Charles I at Oxford; Robert Moray captured by Duke of Bavaria while leading Scots guards; Robert Moray starts correspondence with Kircherus about Magnetism while in Prison in Bavaria
1644	Royalists defeated at Marston Moor by Cromwell.
1645	Robert Moray ransomed for £16,500; Charles I defeated by Parliament in Battle of Naseby; Robert Moray comes to London to negotiate between Charles I, the Scots and the French; Cromwell appointed Cavalry Commander in New Model Army
1646	Charles II (Prince of Wales) flees to Jersey; Charles I flees to Newcastle to the Scots Army; Charles I surrenders to Scottish Army (Covenantors); Fall of Oxford. Prince Rupert surrenders, Ashmole surrenders soon afterwards at Worcester; Parliament sends the Nineteen Propositions to Charles I; Robert Moray tries to arrange for Charles I to escape to France from Newcastle
1647	Charles I goes to Northampton; Charles I handed over to English Parliament; Sir Robert Moray attends the Edinburgh Lodge
1648	Second Civil War starts in Pembroke; Robert Moray goes to France to meet Charles II (then Prince of Wales) to invite him to Scotland; Cromwell defeats Scots at Preston
1649	Charles I put on trial by English Parliament, sentenced to death and executed

Year	*Event*
1650	Charles II agrees to take the Covenant at Breda; Charles II arrives at Speymouth Scotland en route to Aberdeen; Parliament declares war on Scotland; Cromwell's Army crosses the river Tweed into Scotland; Cromwell retreats from Musselburgh to Dunbar; Cromwell returns to Musselburgh; Charles II in Leith; Cromwell falls back to Dunbar; Scots defeated at Dunbar; Charles II publically repents of his and his father's sins, at Perth
1651	Charles II crowned at Scoon and subcribes to the Covenant; Robert Moray made Privy Counsellor in Scotland; Cromwell accepts Chancellorship of Oxford University; Charles II leaves Stirling to march on England; Charles II approaches Border; Charles II defeated at Worcester and flees to France
1652	Cromwell defeats Scots and declares Union in Edinburgh
1653	Robert Moray's wife dies in pregnancy and is buried in Balcarres Fife; Robert Moray goes to Highlands to organise Rising in support of Charles II; Robert Moray opposes Lord Glencairn as leader of Highland Rising for Charles II; Robert Moray imprisoned by Lord Glencairn and accused of plot to assassinate Charles II; Robert Moray writes to Charles II pleading innocence and loyalty addressing Charles a Master Builder
1654	Cromwell dissolves Rump Parliament; Highland Rising led by Glencairn defeated by Cromwell at Battle of Loch Garry
1655	Cromwell allows Jews to return to England; Robert Moray cleared of plot, the letter had been forged, and returns to Paris

Year	*Event*
1656	Robert Moray in Bruges
1657	Robert Moray goes to Maastricht
1658	Oliver Cromwell dies; English Navy captures Dunkerque
1659	Robert Moray was presented to the Masons of Maastricht by Everard Master of the Craft of Masons
1660	General Monck forces the Rump Parliament to dissolve; Charles II invited to return as King of England; Charles II returns to London; William Moray, Robert's younger brother appointed Master of Works and General Warden of Masons in Scotland; Royal Society formally constituted at Gresham College
1661	Robert Moray reappointed Privy Counsellor to Charles II; Robert Moray thanked Charles II in person for the Royal Charter for the Society
1662	Charles II marries Catherine of Braganza; First Charter Granted to Royal Society; Royal Society thanks Robert Moray for his help in obtaining Royal Charter but asks for another one; Charles II grants Royal Society funds
1663	Robert Moray plans demonstration experiments for Charles II's visit to Royal Society; Robert Moray plans a survey of the stars of the Zodiac by Members of the Royal Society; Second Charter Granted to Royal Society; Robert Moray carries out experiments for Charles II
1665	Robert Moray starts to write a history of Freemasonry; Robert Moray notes in letter that he has completed twenty-four pages of his History of Freemasonry

Year	*Event*
1666	Robert Moray writes to John Evelyn, 'It seems you conclude me to be a greater Master in another sort of Philosophy than that which is the businese of the Royall Society'
1667	Robert Moray goes to Scotland as Charles II's commissioner
1669	Sir William Moray resigns as General Warden of Scotland
1670	Charles II uses Masonic identification to Moray at Windsor; Robert Moray quarrels with Lauderdale
1672	England at war with Netherlands
1673	Robert Moray dies and is buried in Westminster Abbey
1685	Charles II dies
1688	Glorious Revolution

Index

G

K

L

M

N

Q